SPRATTON C.E.
PRIMARY SCHOOL

Into the
SPACE AGE

1960 TO 1975

Published by The Reader's Digest Association, Inc.
London • New York • Sydney • Montreal

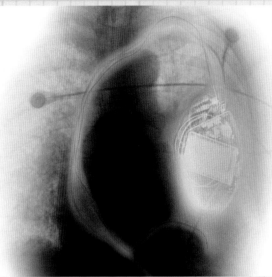

Contents

Introduction	4
THE STORY OF INVENTIONS	18
Communications satellites	20
Genesis of the satellite	26
The Scopitone	28
International system of units	28
The audio cassette	29
The Tetra Pak	29
Laser eye surgery	30
The modem	31
Hang gliders	32
ROBOTICS	36
Shape memory alloys	42
The synthesiser	43
HAROUN TAZIEFF	44
Carbon fibre	46
FRANÇOIS JACOB AND JACQUES MONOD	50
Word processing	52
Acrylic paints	54
Photochromic lenses	54
The mini-skirt	55
Tidal power stations	56
The hovertrain	58
The container ship	59
Pulsars	60
The sailboard	62
The computer mouse	63
Liquid crystal displays	64
Electronic eyes	68
Floppy disks	69
Quarks	70
WERNHER VON BRAUN	72
Apollo 11 mission	74
The ioniser	80
POSM – Patient-Operated Selector Mechanism	80
Implantable defibrillator	81
CIVIL AVIATION	82
The water jet cutter	87
The microprocessor	88
SPACE TRAVELLERS	90
Space stations	94
The cash machine	98
Soft contant lenses	99
The monoski and snowboard	100
The personal computer	102
THE DISPOSABLE SOCIETY	108
The pocket calculator	112
Chaos theory	114
CT scanning	116
MRI scanning	117
The hard disk drive	118
The skateboard	119
Anti-lock braking	119
TRAFFIC IN TOWNS	120
Mountain bikes	124
Rubik's Cube	125
Catalytic converters	125
Smart cards	126
MINIATURISATION	130
The discovery of Lucy	134
Endorphins	136
SILICON VALLEY	138
SNAPSHOT OF SAN FRANCISCO	140
CHRONOLOGY	144
Index	154
Acknowledgments	159

Introduction

The 1960s were an extraordinary time of expanding personal freedom, economic growth and impassioned calls for peace. Politically, the world was separated into two rival power blocs led by the USA and USSR, but once the traumas of the erection of the Berlin Wall (1961) and the Cuban missile crisis (1962) had passed, peaceful coexistence became largely the order of the day. The two superpowers still locked horns on proxy battlefields, fearful that their respective ideologies of capitalism and communism were under threat. Some of these conflicts – in particular the Vietnam War, which involved increasing numbers of US troops from 1964 to 1975 – would prove cruelly destructive and ultimately misguided. Yet for the most part the superpowers competed economically and scientifically, flinging themselves into the great adventure of the space race. The USSR was the early leader, putting both the first satellite and the first man into space, but the rivalry culminated in 1969 with the Apollo 11 Moon mission and a clear victory for the USA.

For the old imperial powers decolonisation was the order of the day, often at a heavy cost. And new political experiments got under way: Che Guevara became an icon for a younger generation eager for change. The tide of hope that accompanied the election of John F Kennedy as US President in 1960 and the civil rights campaign led by Martin Luther King Jr was mirrored by the despair that followed the assassination of both these remarkable men, respectively in 1963 and 1968, and then of Kennedy's younger brother Robert, also in 1968. That same year, the flowers that bloomed in the Prague Spring were crushed under the treads of Soviet tanks. In France the violent events of May 1968 fed into a cultural, but not a political, revolution.

Despite the turbulence on the streets, the mood in the West, encouraged by full employment, was mostly optimistic. Feminists sought new roles for women, sex became more liberated, and young hippies dreamed of creating a utopia as the Beatles sang 'All You Need Is Love'. The Woodstock Festival marked the zenith of this era of idealism, but also the start of its decline. By the early 1970s the climate was already less carefree and more self-questioning. Then in 1973 OPEC, the Organization of Petroleum Exporting Countries, announced an unprecedented hike in world oil prices, shaking an international economy that depended on fossil fuels. It was the end of an enchanted era.

The editors

A New View of Our Planet
This famous photograph showing the Earth rising above the lunar horizon was taken by the crew of the Apollo 8 Moon mission on 24 December, 1968. The image was the first to show the Earth as it looks from deep space and was hugely influential in the environmental movement. Crewed by astronauts Frank Borman, James Lovell and William Anders, Apollo 8 was the first manned space mission to orbit the Moon. The first orbit of the Moon had been achieved in April 1966 by Lunar 10, an unmanned spacecraft launched by the USSR.

▼ Launched by rocket from Cape Canaveral in the US state of Florida, on 10 July, 1962, the Telstar 1 communications satellite relayed the first live transatlantic television transmissions two weeks later

▶ On 31 January, 1958, the Juno 1 booster rocket launched the first American satellite, Explorer 1, into Earth orbit; four months earlier, the successful launch of the world's first satellite, Sputnik 1, had been announced by the USSR

The 1960s saw the emergence of a new breed of hero, as astronauts took their place alongside the explorers of earlier times. Space supplanted the oceans, and the Moon became the new Promised Land. The Soviets launched the first artificial

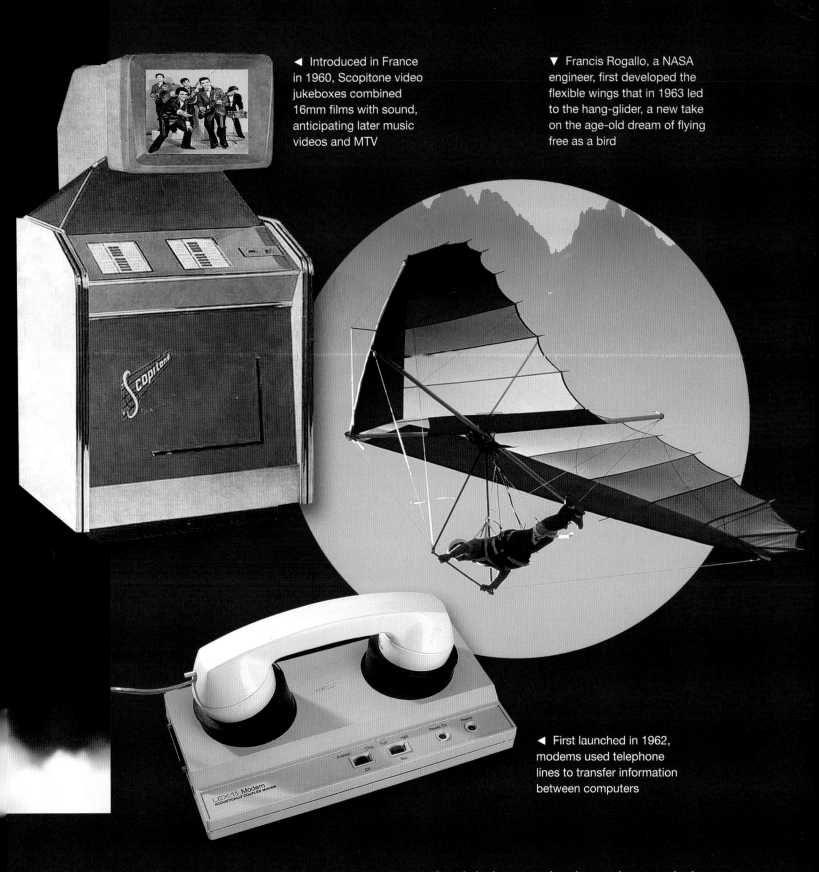

◀ Introduced in France in 1960, Scopitone video jukeboxes combined 16mm films with sound, anticipating later music videos and MTV

▼ Francis Rogallo, a NASA engineer, first developed the flexible wings that in 1963 led to the hang-glider, a new take on the age-old dream of flying free as a bird

◀ First launched in 1962, modems used telephone lines to transfer information between computers

satellites, Sputnik 1 and 2, the latter of which carried on board the dog Laika, the first living being in space. Yuri Gagarin became the first man in space soon after. The US reply culminated in the Apollo 11 mission, which put Neil Armstrong and Buzz Aldrin on

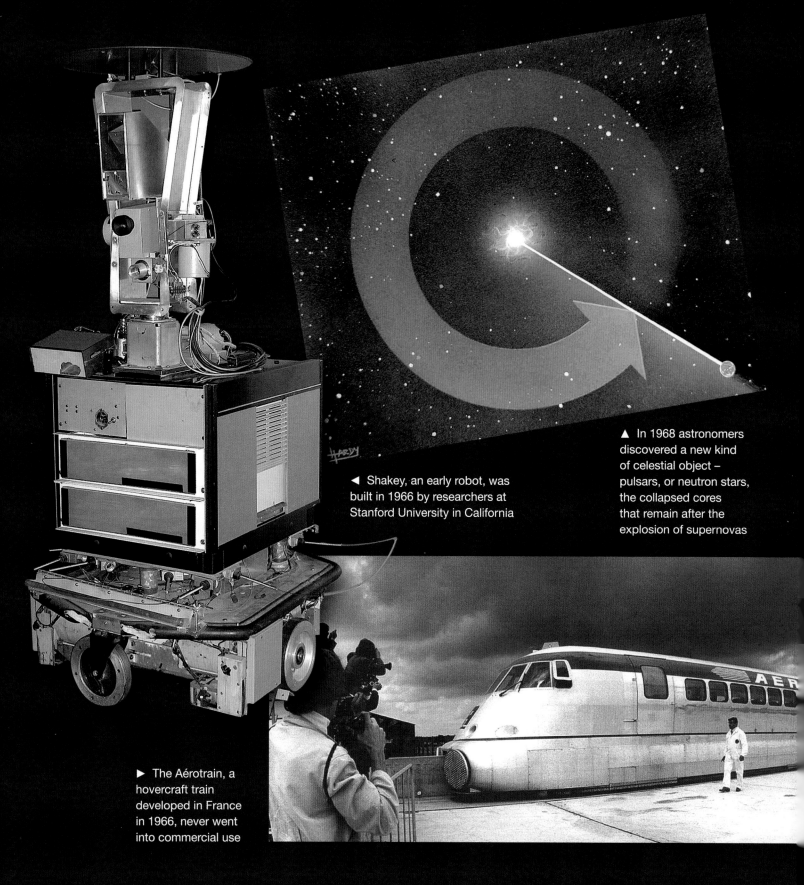

◀ Shakey, an early robot, was built in 1966 by researchers at Stanford University in California

▲ In 1968 astronomers discovered a new kind of celestial object – pulsars, or neutron stars, the collapsed cores that remain after the explosion of supernovas

▶ The Aérotrain, a hovercraft train developed in France in 1966, never went into commercial use

the lunar surface on 21 July, 1969. The USSR responded two years later with the first space station, Salyut 1. Space exploration made a huge contribution to the development of technology, particularly in electronics, which saw microprocessors introduced

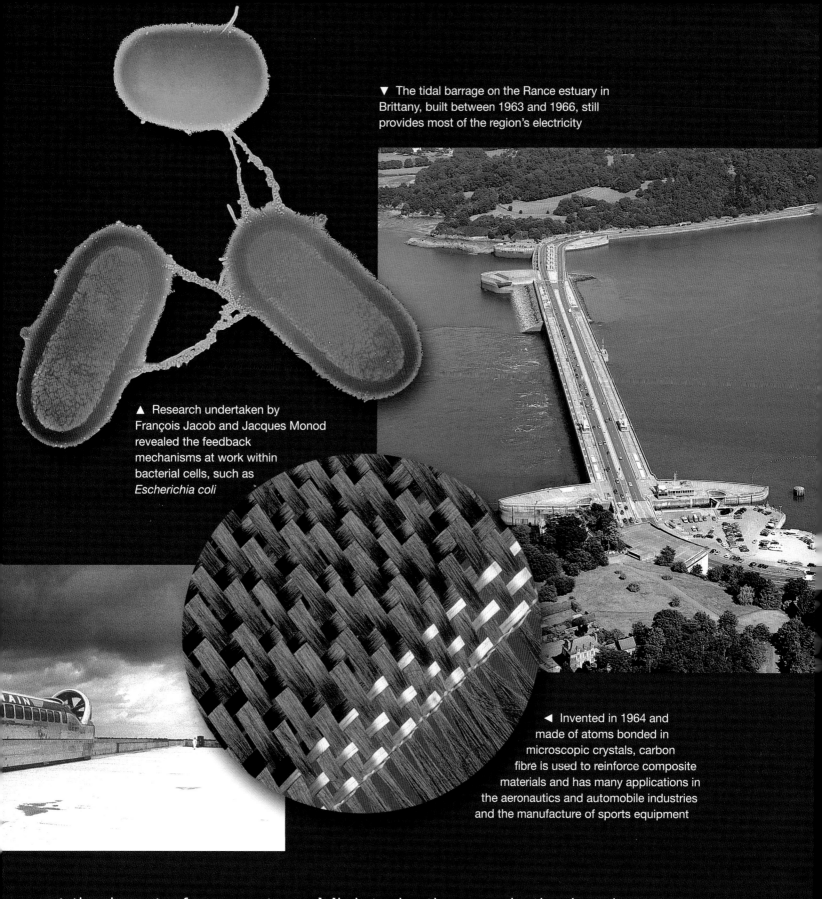

▼ The tidal barrage on the Rance estuary in Brittany, built between 1963 and 1966, still provides most of the region's electricity

▲ Research undertaken by François Jacob and Jacques Monod revealed the feedback mechanisms at work within bacterial cells, such as *Escherichia coli*

◄ Invented in 1964 and made of atoms bonded in microscopic crystals, carbon fibre is used to reinforce composite materials and has many applications in the aeronautics and automobile industries and the manufacture of sports equipment

at the heart of computers. Miniaturisation made the hardware involved smaller and smaller until the first office computers put in an appearance, equipped with hard and floppy disks, modems and the mouse. Meanwhile, watches and calculators were being

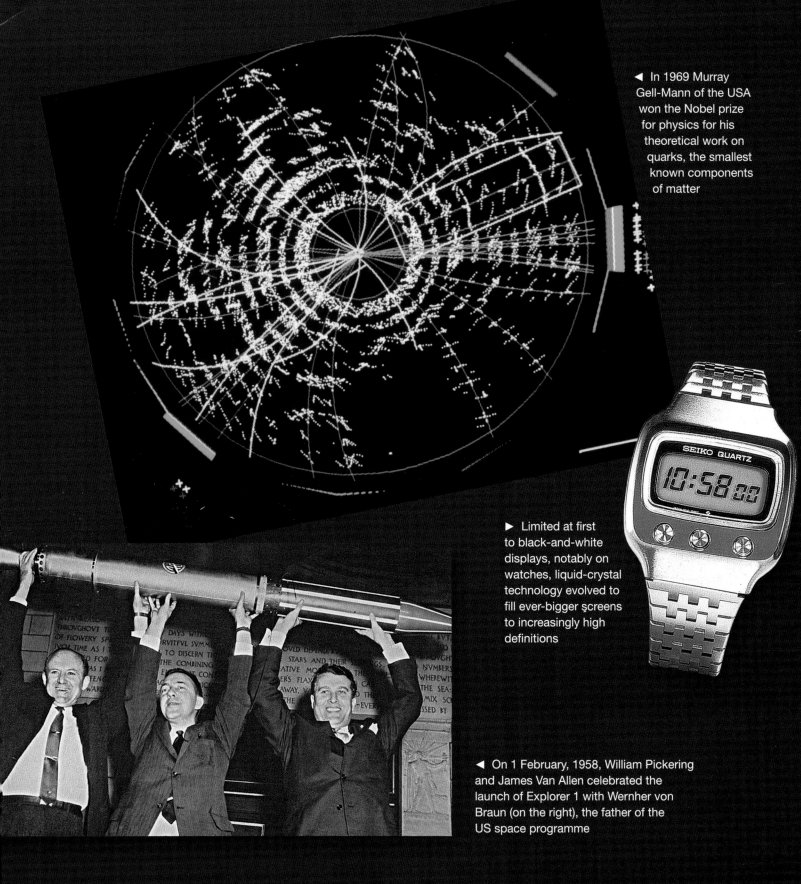

◄ In 1969 Murray Gell-Mann of the USA won the Nobel prize for physics for his theoretical work on quarks, the smallest known components of matter

► Limited at first to black-and-white displays, notably on watches, liquid-crystal technology evolved to fill ever-bigger screens to increasingly high definitions

◄ On 1 February, 1958, William Pickering and James Van Allen celebrated the launch of Explorer 1 with Wernher von Braun (on the right), the father of the US space programme

equipped with liquid-crystal displays (LCDs), and increasingly sophisticated instruments were opening fresh windows on the universe. Radio telescopes scanning the depths of space observed rotating stars called pulsars, while at the other end of

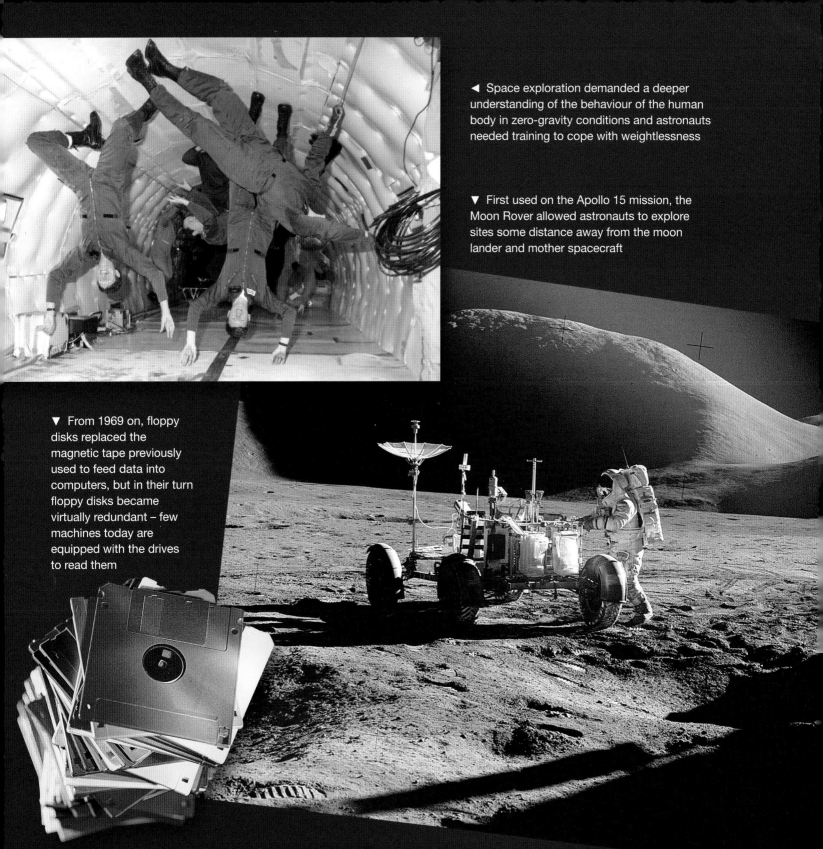

◄ Space exploration demanded a deeper understanding of the behaviour of the human body in zero-gravity conditions and astronauts needed training to cope with weightlessness

▼ First used on the Apollo 15 mission, the Moon Rover allowed astronauts to explore sites some distance away from the moon lander and mother spacecraft

▼ From 1969 on, floppy disks replaced the magnetic tape previously used to feed data into computers, but in their turn floppy disks became virtually redundant – few machines today are equipped with the drives to read them

the size spectrum particle accelerators revealed the existence of quarks, the smallest constituent particles of protons and neutrons. Volcanologist Haroun Tazieff, among others, cast fresh light on the workings of planet Earth, and in 1972 the US meteorologist

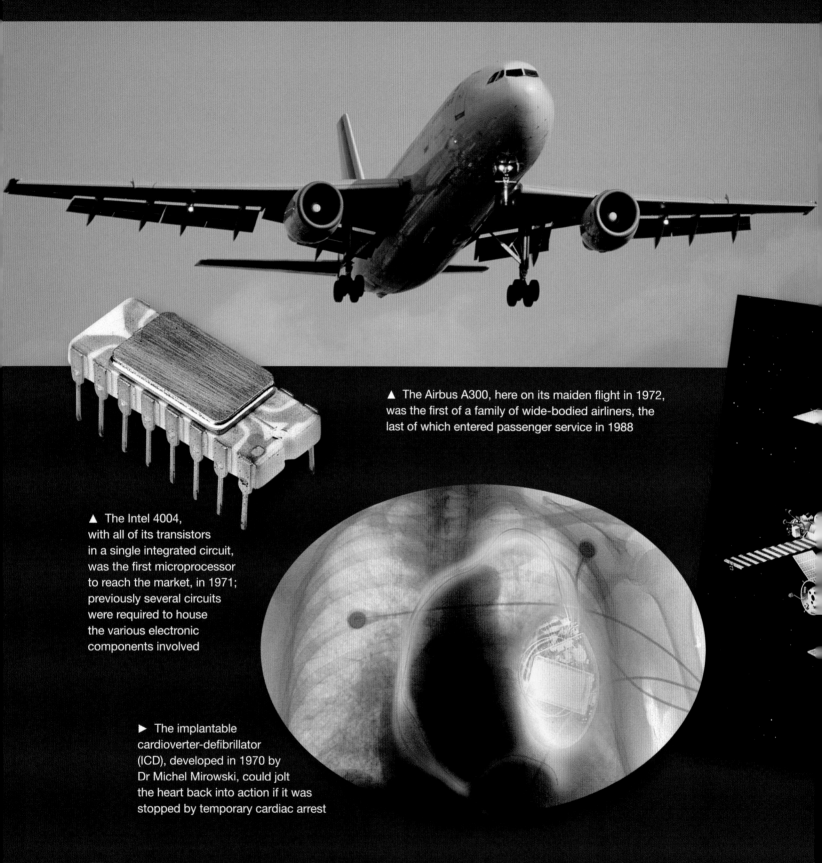

▲ The Airbus A300, here on its maiden flight in 1972, was the first of a family of wide-bodied airliners, the last of which entered passenger service in 1988

▲ The Intel 4004, with all of its transistors in a single integrated circuit, was the first microprocessor to reach the market, in 1971; previously several circuits were required to house the various electronic components involved

▶ The implantable cardioverter-defibrillator (ICD), developed in 1970 by Dr Michel Mirowski, could jolt the heart back into action if it was stopped by temporary cardiac arrest

Edward Lorenz hosted a conference on climate that marked the birth of chaos theory. Technology also made major advances in the leisure field. Synthesisers brought new sounds to music, which was more widely diffused than ever thanks to audio cassettes and

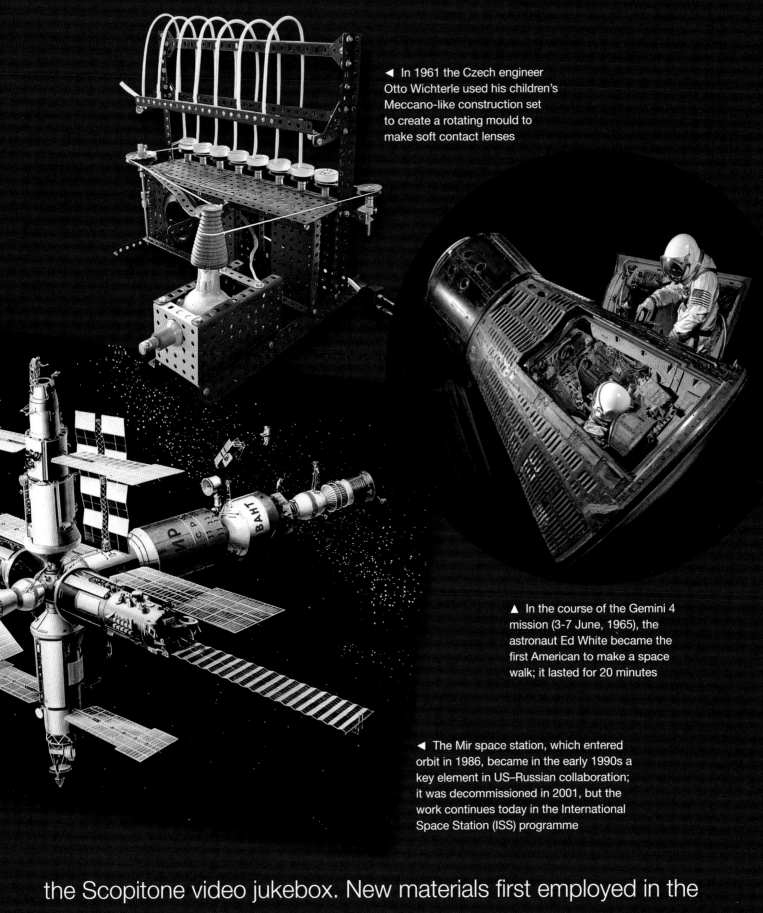

◀ In 1961 the Czech engineer Otto Wichterle used his children's Meccano-like construction set to create a rotating mould to make soft contact lenses

▲ In the course of the Gemini 4 mission (3-7 June, 1965), the astronaut Ed White became the first American to make a space walk; it lasted for 20 minutes

◀ The Mir space station, which entered orbit in 1986, became in the early 1990s a key element in US–Russian collaboration; it was decommissioned in 2001, but the work continues today in the International Space Station (ISS) programme

the Scopitone video jukebox. New materials first employed in the aeronautics industry, such as carbon fibre, were put to use in a new generation of sports equipment, from windsurfers and snow boards to tennis racquets and mountain bikes. Research on genes

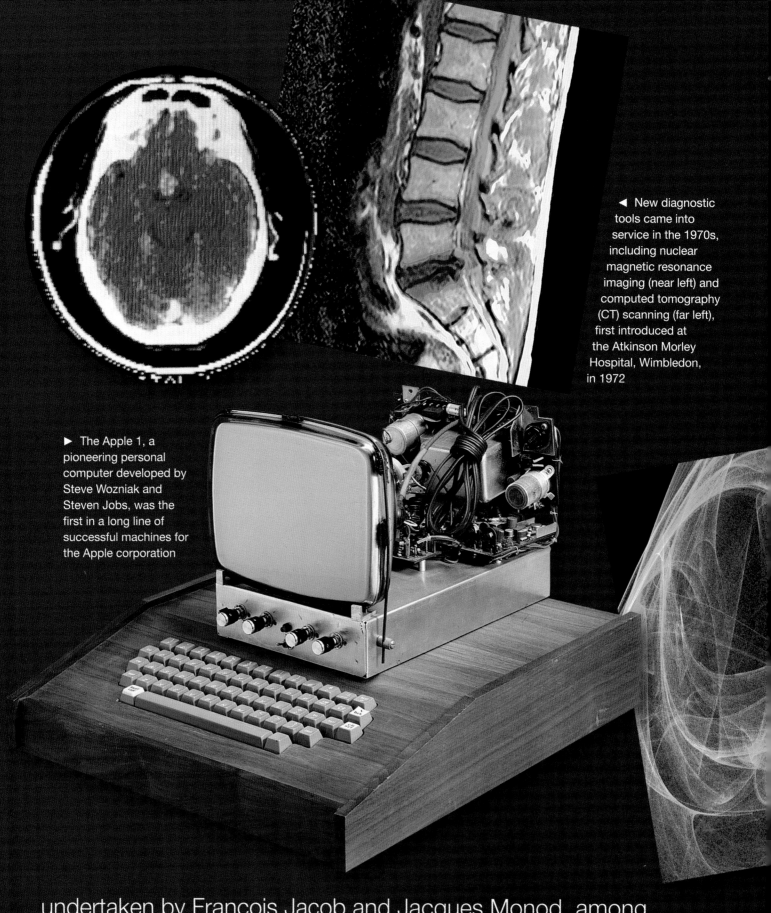

◄ New diagnostic tools came into service in the 1970s, including nuclear magnetic resonance imaging (near left) and computed tomography (CT) scanning (far left), first introduced at the Atkinson Morley Hospital, Wimbledon, in 1972

► The Apple 1, a pioneering personal computer developed by Steve Wozniak and Steven Jobs, was the first in a long line of successful machines for the Apple corporation

undertaken by François Jacob and Jacques Monod, among others, provided a better understanding of the workings of genetic inheritance. Hospitals benefited from a new range of intelligent machines – scanners, magnetic resonance imagers, implantable

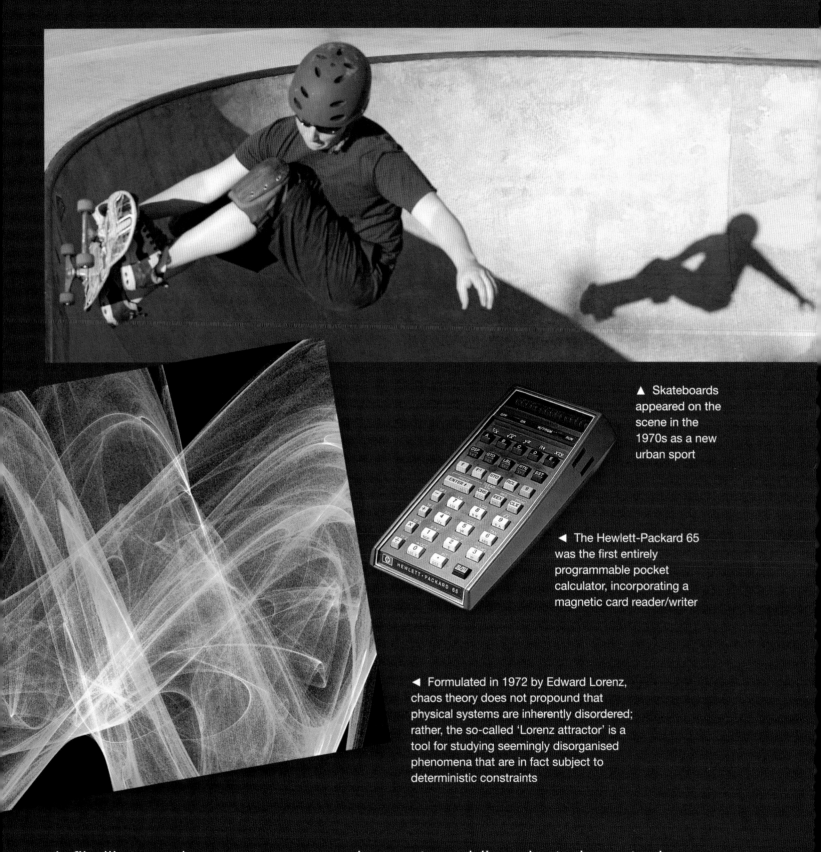

▲ Skateboards appeared on the scene in the 1970s as a new urban sport

◄ The Hewlett-Packard 65 was the first entirely programmable pocket calculator, incorporating a magnetic card reader/writer

◄ Formulated in 1972 by Edward Lorenz, chaos theory does not propound that physical systems are inherently disordered; rather, the so-called 'Lorenz attractor' is a tool for studying seemingly disorganised phenomena that are in fact subject to deterministic constraints

defibrillators, laser surgery equipment – while robots boosted productivity in industry. In a time of double-digit economic growth, trade expanded partly thanks to the advent of containers and container ships. Aviation benefited from the spread of

▼ Invented by Roland Moreno, smart cards could store a large amount of data in an integrated circuit – credit cards and phone cards evolved from this revolutionary concept

▲ The Place de la Concorde in Paris clogged with traffic in the early 1960s, suggesting the changes that would soon affect major cities now that the car was king

◄ Staff at Fairchild Semiconductors in California's Silicon Valley, an electronics corporation that marketed the first integrated circuits from the late 1950s on

jet propulsion and the advent of high-speed airliners helped to democratise travel. Mobility became the new credo, encouraged by meteoric growth in car ownership, which began to create serious traffic congestion problems from the early 1970s on.

◀ Miniaturisation revolutionised the IT (information technology) industry – the large hard disk dates from 1984, the small one from 1999

▼ Endorphins are hormones produced by the brain that explain the differing sensitivity of individuals to pain, stress and euphoric pleasure

◀ San Francisco became the symbol of the counter-culture and of psychedelic rock, which could be heard at the Matrix and the Fillmore Auditorium from 1965, at the Avalon in 1966 and at the Fillmore West two years later

Two locations in California summed up the spirit of the age: Silicon Valley, where information technology was born, and the city of San Francisco, which won international fame as a magnet for hippies, musicians and the gay community.

THE STORY OF INVENTIONS

At the start of the 1960s human beings acquired new tools that extended their capabilities. Artificial satellites and space exploration gave us a fresh view of the Earth – seen from space, the blue planet was revealed in all its beauty and fragility. Meanwhile robotics was gradually reducing the demand for repetitive manual labour and replacing the force of nature with that of machines. A whole new world of possibilities was opening up.

COMMUNICATIONS SATELLITES – 1960

The Earth seen from above

Born of the rapid progress made in missile technology in the Second World War, artificial satellites colonised near-Earth space in the late 1960s. Besides allowing humans to see their own planet from a distance for the first time, these 'eyes in the sky' were soon performing a range of other functions.

On 4 October, 1960, the National Aeronautics and Space Administration (NASA) – barely two years old at the time – launched the world's first communications satellite, the Courier 1-B. The commercial exploitation of space had begun. In the 1960s alone, the US and Soviet space programmes launched 45 separate satellites. The inner reaches of space were becoming a suburb of the planet.

The Cold War in space

Three years to the day earlier, on 4 October, 1957, an aluminium sphere measuring 58cm in diameter and weighing 84kg entered history under the name of Sputnik 1 (the word means 'satellite' in Russian). Put in low Earth orbit, it sent back *beep-beep* radio signals that transfixed the world. The exploration of space was under way, and the Soviets were first off the starting block.

To achieve the feat of launching Sputnik the Soviet scientists had transformed an R7 intercontinental missile into a space rocket. The R7 was already capable of travelling at 25,000km/h (15,500mph). Equipped with a booster, it could reach speeds of 28,000km/h (17,400mph), the minimum velocity required for a rocket to escape the Earth's gravitational pull and go into orbit.

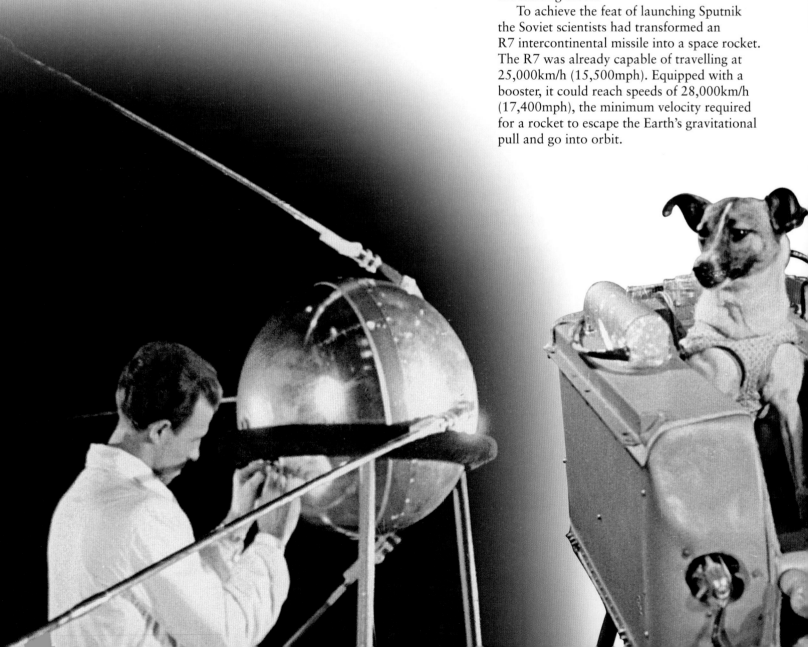

Early failure
The USA's first attempt to launch a satellite ended badly when the Vanguard TV-3 launch rocket exploded on take-off at Cape Canaveral on 6 December, 1957 (far right). Thrown clear, the satellite was found intact a few metres away. The story was front-page news across the nation as newspapers dubbed the disaster 'Kaputnik' (right).

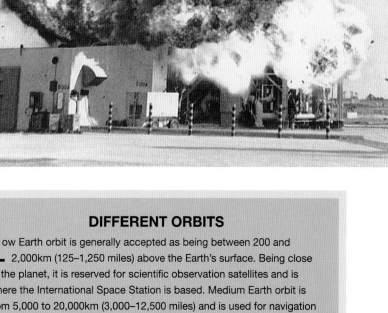

Sputnik 2 was launched in November 1957, carrying Laika, a dog that became the first creature to live (and die, probably from overheating) in space. Manned spaceflight was becoming a possibility.

The USA was quick to respond. The Explorer 1 satellite went into orbit on 31 January, 1958, placed there by a Redstone short-range ballistic missile that, like the R7, had been adapted for the purpose. Like the Sputniks, Explorer 1 was a scientific research satellite designed to study conditions in the

First moves in space
The Soviets made the running in the early days of the space race, as celebrated in a documentary film of 1969 which showed scientists working on Sputnik 1 (far left), as well as the sad fate of Laika (left), a mongrel bitch fired into space on Sputnik 2 to see if life was possible in space conditions.

DIFFERENT ORBITS

Low Earth orbit is generally accepted as being between 200 and 2,000km (125–1,250 miles) above the Earth's surface. Being close to the planet, it is reserved for scientific observation satellites and is where the International Space Station is based. Medium Earth orbit is from 5,000 to 20,000km (3,000–12,500 miles) and is used for navigation satellites along the GPS model. Geostationary orbit, about 36,000km (22,000 miles) above the Equator, is favoured for telecommunications and weather satellites. Bodies placed there move in synchronisation with the Earth's own rotation, so that from Earth they appear to be stationary in the sky. Polar orbit passes over the North and South Poles at a height of 400 to 1,000km (250–620 miles). As the satellite passes from one pole to the other, the Earth rotates beneath it, revealing separate strips of its surface that can be assembled to form a composite picture. In this way a single polar satellite can build up an image of the whole planet.

COMMUNICATIONS SATELLITES – 1960

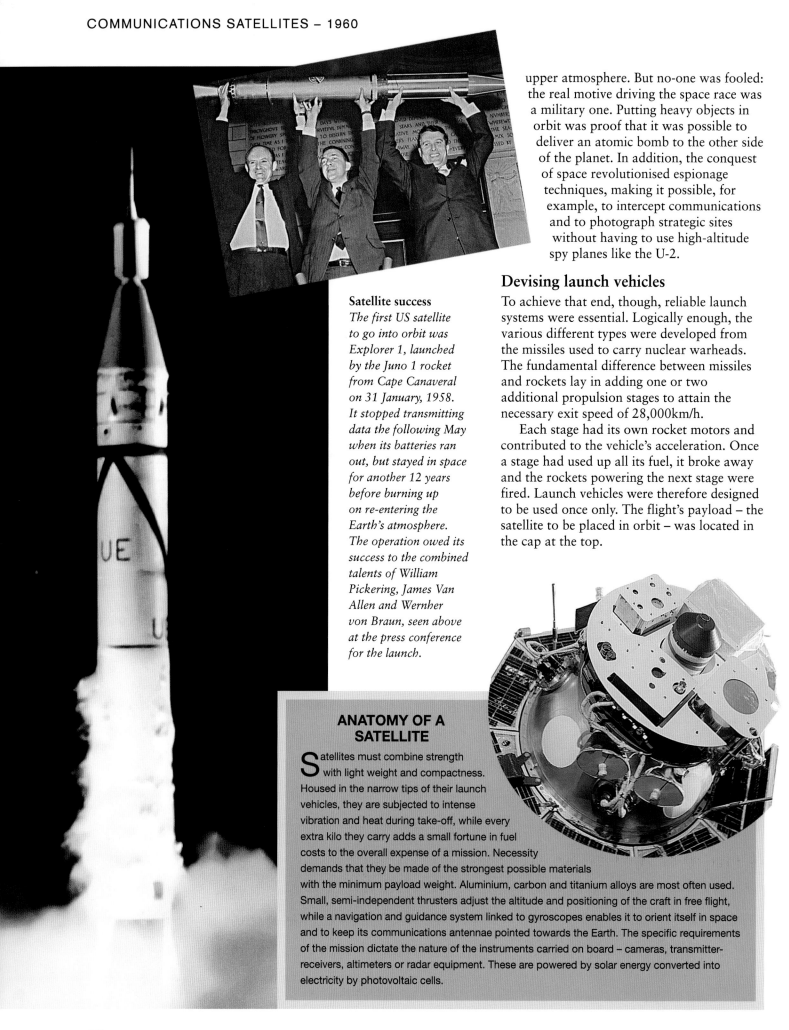

upper atmosphere. But no-one was fooled: the real motive driving the space race was a military one. Putting heavy objects in orbit was proof that it was possible to deliver an atomic bomb to the other side of the planet. In addition, the conquest of space revolutionised espionage techniques, making it possible, for example, to intercept communications and to photograph strategic sites without having to use high-altitude spy planes like the U-2.

Devising launch vehicles

To achieve that end, though, reliable launch systems were essential. Logically enough, the various different types were developed from the missiles used to carry nuclear warheads. The fundamental difference between missiles and rockets lay in adding one or two additional propulsion stages to attain the necessary exit speed of 28,000km/h.

Each stage had its own rocket motors and contributed to the vehicle's acceleration. Once a stage had used up all its fuel, it broke away and the rockets powering the next stage were fired. Launch vehicles were therefore designed to be used once only. The flight's payload – the satellite to be placed in orbit – was located in the cap at the top.

Satellite success
The first US satellite to go into orbit was Explorer 1, launched by the Juno 1 rocket from Cape Canaveral on 31 January, 1958. It stopped transmitting data the following May when its batteries ran out, but stayed in space for another 12 years before burning up on re-entering the Earth's atmosphere. The operation owed its success to the combined talents of William Pickering, James Van Allen and Wernher von Braun, seen above at the press conference for the launch.

ANATOMY OF A SATELLITE

Satellites must combine strength with light weight and compactness. Housed in the narrow tips of their launch vehicles, they are subjected to intense vibration and heat during take-off, while every extra kilo they carry adds a small fortune in fuel costs to the overall expense of a mission. Necessity demands that they be made of the strongest possible materials with the minimum payload weight. Aluminium, carbon and titanium alloys are most often used. Small, semi-independent thrusters adjust the altitude and positioning of the craft in free flight, while a navigation and guidance system linked to gyroscopes enables it to orient itself in space and to keep its communications antennae pointed towards the Earth. The specific requirements of the mission dictate the nature of the instruments carried on board – cameras, transmitter-receivers, altimeters or radar equipment. These are powered by solar energy converted into electricity by photovoltaic cells.

THE PROBLEM OF SPACE DEBRIS

Since the start of the Space Age the Earth's near environment has been transformed into a giant dumping-ground littered with abandoned rocket stages, decommissioned satellites, junk metal and much else. Some 13,000 objects more than 10cm in diameter, and a further 200,000 measuring between 1 and 10cm, continually circle the planet. Given that each object is travelling at about 30,000km/h (18,500mph), even the smallest nut or bolt can be as lethal as a bullet. To protect astronauts and satellites, all pieces of debris bigger than 10cm are tracked by radar so that craft can avoid their trajectories. The only way of dealing with the rest is armour-plating – spacecraft are reinforced to increase their resistance. For a definitive solution, space agencies have proposed using a space shuttle or a specially designed satellite to force most of the objects back towards the Earth, where they would burn up on re-entry into the atmosphere. But any such operation would be expensive, so the prospect remains uncertain.

Final adjustments
A technician carries out checks on Asterix, the first French satellite. It had no scientific instrumentation on board, being solely intended to test the performance of the Diamant launch rocket.

The Atlas-Centaur, used from 1962 on, could put objects weighing up to 1.5 tonnes in orbit. Thor Deltas were less powerful, carrying a maximum load of 250kg. The Titan rockets, derived from Titan 2 missiles, were the most powerful of all. In service from 1966, they could place 5-tonne satellites in geostationary orbit and ones weighing up to 17 tonnes in low Earth orbit.

In the USSR, the Semyorka rocket was developed from the R7, the first in a long line of launch vehicles that is still operational today. Even the very first, which put the Sputnik satellites in space, could carry payloads up to 5 tonnes. Other versions adapted for different types of mission were introduced in the course of the 1960s. The C model (1964) was used for light military or scientific satellites in low Earth orbit. The DS (1965) had a carrying capacity of 19 tonnes and serviced geostationary orbits. They would later be used to deliver modules to the Salyut and Mir manned space stations.

By the mid 1960s the two superpowers had a range of reliable launch systems for different types of satellite and varying orbits. Launches became more frequent: there were 65 in 1962, three years later there were 180. On 26 November, 1965, France became the third nation to venture into space when it put the A1 Asterix satellite into orbit. Weighing 38kg, it was carried by a Diamant rocket launched from Hammaguir in the Algerian Sahara. From 1968 on, Kourou in French Guiana became France's principal launch site.

Mapping the debris
The visualised image above shows the accumulation of litter in low Earth orbit – the size of the objects has been exaggerated to make them visible. The most significant part of the debris consists of almost 2,000 defunct satellites that still circle the planet.

COMMUNICATIONS SATELLITES – 1960

Spy in the sky
The synthetic image below shows the Helios 2B military reconnaissance satellite, launched on 18 December, 2009, by an Ariane rocket. The programme is supported by France, Belgium, Spain, Italy and Greece.

A range of uses

In the 1960s and 70s, Earth orbit filled up with satellites of various kinds dedicated to various uses, from telecommunications and scientific research to spying. Whatever their function, all satellites were very much of a kind. Slipping silently through the heavens, they dramatically affected the lives of the humans below them.

From the 1960s on, US reconnaissance satellites – first SAMOS ferrets, then Chalets – spied on Soviet and Chinese communications. The USSR responded with EORSAT devices.

The spy satellites took vast numbers of high-resolution photographs of sensitive sites, gathering information on everything from troop movements to the activity of industrial plants. Initially they could provide clear images of objects up to 200 metres across; today they can zoom in to within just a few metres.

Meanwhile, commercial satellites were revolutionising telecommunications. Usually set in geostationary orbit, these received signals from one or more transmission stations situated on Earth and relayed them over great distances, including between continents. The first live television link between the USA and Europe was established on 11 July, 1962, by Telstar 1, developed by the AT&T corporation and its subsidiary Bell. In 1965 Intelsat 1, also known as Early Bird, launched the era of satellite telephone communication.

Satellites also had a profound effect on navigation on land and at sea by making geolocation from space a practical possibility. Science was another winner: astronomers were given a much clearer view of deep space after 1990, when NASA put the Hubble space telescope in orbit. Perched 800km (500 miles) above the Earth, it provided a vision of the heavens unimpeded by the perturbations of the planet's atmosphere.

From their high vantage point satellites scrutinised the Earth as never before, casting new light on climate, wind circulation and sea levels. Weather forecasting was made

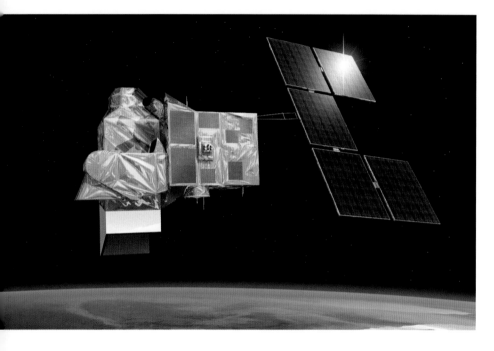

WEATHER SATELLITES

Since the launch of TIROS 1 in 1960, weather satellites have continually observed the Earth, providing meteorologists with a huge amount of data on wind direction, the formation of cloud banks, temperature and the distribution of rainfall. Previously weather forecasters relied on information from Earth stations, which did not provide global coverage. Successive generations of Landsat, SPOT and Meteosat devices record all these parameters with ever greater precision. Processed by increasingly sophisticated IT equipment – notably the supercomputers introduced in the 1980s – the data enabled meteorologists to devise models predicting changes in the atmosphere and to give weather forecasts up to 10 days in advance.

Trouble ahead
An image captured by the TOPEX/Poseidon weather satellite shows the El Niño climate phenomenon on 10 December, 1997. The raised temperature of the Pacific Ocean (shown in fuchsia and white) had disastrous consequences for the entire region.

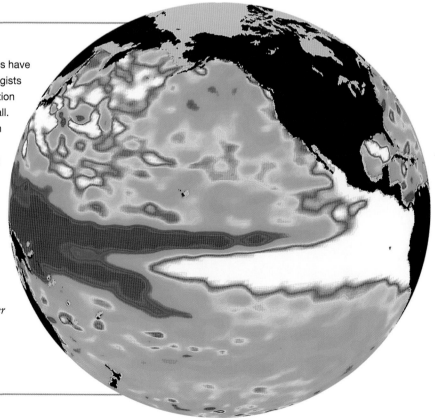

Real-time eruption
A SPOT Earth-observation satellite captures volcanic activity from Mount Etna on Sicily. The programme came into service in 1986 and now employs five satellites equipped with state-of-the-art imaging equipment, the most up-to-date of which provides a resolution of 2.5–5m.

Finishing touches
Technicians prepare the telecommunications satellite Horizons-2 for launch at the European Space Agency's base at Kourou, French Guiana, in December 2007.

more accurate, and from the 1960s on satellites were vital tools in keeping watch on the warming of the planet.

Space activity today

The USA has the world's foremost presence in space, accounting for 75 per cent of the global space budget: in 2007 it invested $57 billion. France came next, with a budget equivalent to $2.9 billion, then Japan with $2.2 billion. By that time attention had turned to cutting the cost of space launches and prolonging the life of satellites, which currently average 5 years but it is hoped will rise to 10 then 15 years.

Two more nations have taken up the space baton. China launched its first satellite in 1970 and its first manned space flight in 2003; it already has distant plans for a manned lunar landing. India has been launching satellites successfully since 1975 and has 60 space missions planned by 2013.

GEOLOCATION FROM SPACE

In 1964 American scientists introduced TRANSIT, which involved six satellites permanently covering the globe to provide location data accurate to within 200m. The system remained in use until 1996, when it was replaced by the GPS with its fleet of 24 satellites. Geolocation from space has many applications, both civil – marine navigation, for example – and military, such as helping troops find their way in enemy territory. In the 1980s the Soviets developed the GLOSNASS network, while the European Union is currently developing Galileo. Today's systems are accurate to within a few metres.

GENESIS OF AN INVENTION
Communications satellite

The development of electronics and the emergence of terrestrial broadcasting were the first steps in making communications satellites possible. Another prerequisite was the advent of launch vehicles to put them into orbit; these evolved from Second World War missiles.

Helios 2-A military reconnaissance satellite

LAUNCH VEHICLES
FROM NEWTON TO VON BRAUN

In 1687 the physicist Sir Isaac Newton, in his *Principia Mathematica*, expounded the three laws of motion, the third of which – that of action and reaction – is the basic theory behind rocket propulsion. In 1903 Russia's Konstantin Tsiolkovsky established a blueprint for rocketry in his book *The Exploration of Cosmic Space by means of Reaction Devices*, in which he calculated the escape velocity necessary for rockets to leave Earth's atmosphere and demonstrated that multi-stage vehicles would be needed to attain it. In 1932, on the other side of the Atlantic, Robert Goddard constructed the first liquid-propellant rocket. Four years later a Frenchman, Louis Damblanc, built the first two-stage, then three-stage devices. The V-1 and V-2 missiles developed in Germany by Wernher von Braun's team in the Second World War also speeded up the development of launch vehicles; after the war both the Soviets and the Americans built on his work to produce their own prototypes. Since then the weight of the devices launched by rockets into space has risen from 80kg to over 10 tonnes.

Thor-Delta launch vehicle

SOLAR PANELS
FROM SELENIUM TO SILICON

In 1839 the French scientist Antoine Becquerel discovered that light can generate an electric current in certain materials. In the 1870s the first selenium photoelectric cells were produced as a result of the work of Willoughby Smith, an English electric engineer, among others, but photoelectricity remained little more than a curiosity until after the Second World War. In 1954 Calvin Fuller and Gerald Pearson, working for AT&T in the USA, developed a diode made of silicon, the principal element in solar panels. It was soon apparent that the Sun was the ideal energy source for satellites, and by 1958 US engineers were using silicon solar cells with an 8 per cent yield to power the devices. Today 14 to 19 per cent is typical, but 40 per cent efficiency has been achieved.

Franco-American TOPEX-Poseidon satellite equipped with solar panels

Telstar 1, the first communications satellite

COMMUNICATION WITH EARTH
FROM COHERERS TO TRANSISTORS

In 1887 the German physicist Heinrich Hertz showed that electromagnetic waves travelling at the speed of light can generate electricity. The first instrument designed to detect the waves was the coherer, invented by French physicist Édouard Branly in 1890. Russia's Alexander Popov improved the device in 1895 by adding a vertical antenna that provided better reception. In 1904 the English physicist John Ambrose Fleming, working from Edison's light bulb, invented the vacuum tube, which proved more sensitive than Branly's coherer. Thanks to the work of US inventor Lee De Forest, diodes became triodes after 1906, further improving sensitivity and allowing extra amplification. From the 1950s on, transistors made it possible to manufacture transmitters and receivers that could communicate on different wavelengths between Earth and space. The first satellite, Sputnik I, was equipped with two single-watt radio transmitters powered by a couple of silver–zinc batteries that emitted signals for 22 days – longer than the 14 days originally expected.

Antenna dish at the Goonhilly Downs Satellite Earth Station in Cornwall

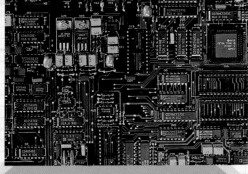

Circuit board for equipment fitted in a satellite

ON-BOARD ELECTRONICS
FROM CALCULATORS TO COMPUTERS

In 1937 the English mathematician Alan Turing outlined the principle of a 'universal machine', establishing the concept of programmes and programming. The first computers, developed in Britain and the USA during the war, were cumbersome because of their use of vacuum tubes. The discovery of transistors by William Shockley and colleagues in 1947 considerably shrank their size. With Jack Kilby's invention of integrated circuits in the USA in 1958 and then the development of microprocessors from 1971 on, electronics was able to provide satellites with 'brains' capable of monitoring such on-board activities as temperature, electricity supply and the direction in which the antennae were pointing. As computer science progressed, satellites became increasingly self-controlling, for example becoming equipped to fire micro-rockets to correct their trajectory – an important function given the need to keep communications antennae pointed towards receivers on Earth.

THE FUTURE OF SATELLITES
SMALLER AND SMARTER

Tomorrow's satellites will be smaller but more effective than those that have gone before. Progress in information technology and miniaturisation will permit the development of devices no more than 1m in diameter that yet contain all the equipment necessary to carry out sophisticated missions. Fleets of interlinked mini-satellites will further advance the progress of telecommunications. Similarly, dozens of small devices working together to produce combined images will form the super-telescopes of the future, producing panoramic views of the cosmos at previously unimagined levels of optical resolution.

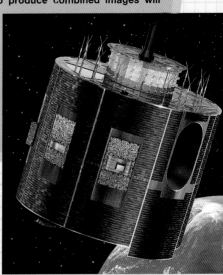

Artist's impression of a second-generation Meteosat weather satellite

INVENTIONS THAT CHANGED OUR LIVES

The Scopitone 1960

Music machine
For small change, Scopitone machines provided video on demand for a public that had not yet had exposure to music videos and for whom television coverage of pop music was still a novelty.

JUKEBOXES YOU COULD WATCH

Well-known artists made films for the Scopitone machines, not just in France where they originated but also in Britain and the USA. Procol Harum sang 'A Whiter Shade of Pale', Nancy Sinatra high-stepped through 'These Boots are Made for Walking' and Neil Sedaka performed 'Calendar Girl'. The films also provided opportunities for young film enthusiasts to hone their skills, among them Claude Lelouch who later directed the Oscar-winning *A Man and a Woman*.

The prototype Scopitone video jukebox, the Scopzione, was first unwrapped by the Italian firm of Cinebox in Milan in 1959. Its commercial successor, the Scopitone, was unveiled the following April in France. An engineer named Frédéric Mathieu developed it for a firm called Cameca, of which he was a director. The machine played pop songs while simultaneously showing films of the artists performing them, allowing fans to see their idols in colour for the first time. Installed in bars, the huge machines stood 2m high and had a 54cm (21-inch) screen. They offered a choice of 30 short films – which were frequently changed – shot in 16mm format with an average playing time of 2 minutes 45 seconds to match the songs. The performances were mimed to the recorded versions of the songs, and the sound was transmitted via a magnetic strip on the film.

Cameca pulled the plug on the venture in 1974, largely as a result of the growing competition from colour television. By that time almost 700 films had been made. Today the Scopitone is remembered, if at all, as a precursor of MTV and the music video.

International system of units 1960

In 1960 the 11th General Conference on Weights and Measures adopted the International System of Units, often abbreviated to SI from its French acronym. Up to that time measurement systems varied from country to country. The SI adopted seven base units: the metre (length), kilogram (weight), second (time), ampere (electric current), kelvin (thermodynamic temperature), mole (amount of substance) and candela (luminous intensity). Each could be multiplied or divided by factors of ten to create other units – centimetres from metres, for example. Scientific communities worldwide adopted this metric system, although some countries – Britain and the USA among them – retained some or all of their old systems for day to day purposes.

Making music
The affordability of audio cassettes made them an important medium for the global diffusion of music. Here, Masai men dance to tapes at Narok in Kenya.

The audio cassette 1961

Invented in 1961, the compact audio cassette was commercialised by Philips of the Netherlands, and introduced by them at the Berlin Radio Show in 1963. Although sound quality was not as good as that from vinyl discs, the devices nonetheless proved popular thanks to their convenient small size. Housed in a plastic case, a magnetic tape wound between two reels, providing up to 90 minutes of stereo recording. At the same time Philips also launched the E1-3-300 cassette player, which was an immediate success, the more so when the firm decided to licence the format free of charge.

In the 1970s cassettes were the only rival to vinyl discs and they claimed almost half the market for recorded music. As an added bonus, they allowed people to record material themselves, either from records that they owned or their own playing. From 1979 onwards, the invention of the Walkman gave cassettes a fresh fillip.

One drawback of the format was the slow pace of the tape – just 4.75cm per second. This limited the sound quality, particularly at high frequencies. Improvements, notably the Dolby systems, were introduced to reduce distortion and background noise. The arrival of compact discs in the early 1980s failed to kill off the audio cassette, but digital recording and downloading finally did so. Attempts to introduce digital cassettes – DAT, DCC and MiniDiscs – proved commercial failures.

> **THE MICROCASSETTE**
>
> Developed by Olympus in 1969, these tiny devices were at first limited by their poor sound quality to use in Dictaphones, then were adopted in telephone answering machines. They found a ready market in those fields until the early 2000s, when they were largely replaced by better-performing digital equivalents.

The Tetra Pak 1961

First proposed by the Swedish industrialist Ruben Rausing in 1951, Tetra Paks were the world's first aseptic drinks cartons. The linings of aluminium and polyethylene meant that perishable liquids could be preserved in the air-tight packs for months on end. The first long-life milk cartons went on the market in 1961, their contents sterilised to ensure that they contained no harmful bacteria. By 2007 more than 137 billion cartons had been sold around the world. By that time manufacturers were increasingly encouraged to use recyclable materials in their production.

INVENTIONS THAT CHANGED OUR LIVES

Laser eye surgery 1961

In 1960 the US physicist Theodore Maiman gave the world the first ruby-crystal laser. The instrument produced a concentrated beam of red light and attracted the attention of ophthalmologists. Carrying intense energy, the ray could be focused on a single point. In New York, Charles Koester and Charles Campbell used the heat produced to burn out a retinal tumour in December 1961. It was the first time lasers had been used for photocoagulation.

Applications and techniques

In 1963 argon laser photocoagulation was used to repair retinal tears. The technique, which is still employed today, could also remove retinal lesions, for example, in diabetic retinopathy. YAG lasers (the letters of the acronym stand for yttrium aluminium garnet, a synthetic crystalline material) were used in the ensuing years to make incisions in ocular membrane. The very brief pulses emitted by YAG lasers, more powerful than those of argon gas lasers, served in effect as surgical knives, allowing them to be used to make incisions in the iris to reduce pressure in cases of glaucoma.

Excimer lasers, with an ultraviolet beam, are used to remodel the surface of the eye to cope with refractive conditions like myopia (shortsightedness), hyperopia (longsightedness) and astigmatism by adjusting the curvature of the cornea to correct the convergence of light rays on the retina. Photorefractive keratectomy (PRK), first described in 1983, is a two-stage operation: first the epithelium, or outer layer of the cornea, is removed, then a computer-guided excimer laser is used to change the shape of the central cornea. An alternative procedure known as LASIK – Laser-Assisted In-situ Keratomileusis – has been used since the early 1990s. Instead of peeling away the epithelium, this involves cutting a hinged flap that is later replaced to protect the remodelled area of the cornea. Both techniques are popular with patients attracted by the convenience and comfort of no longer having to wear spectacles.

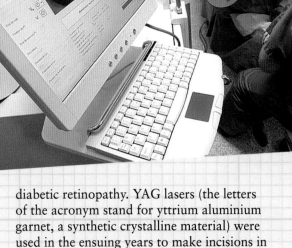

Lasers at work
Surgeons in a hospital in Marseille use a femtosecond laser to correct myopia and astigmatism in a patient.

PUTTING THE LASER IN LASIK

The LASIK procedure was originally performed using a bladed instrument called the microkeratome to cut the corneal flap. Since 2001 this has been replaced by femtosecond lasers, which operate in the infrared frequency by a series of pulses of miniscule duration (one femtosecond equates to a quadrillionth of a second).

Correcting astigmatism
The graphic images show three stages in the correction of astigmatism using the original LASIK method. The condition is caused by a deformation of the cornea, which makes it oval rather than round. The operation involved cutting a flap (top) in the cornea's outer layer with the aid of a microkeratome. An excimer laser was then used to restore the spherical shape of the deep cornea (centre), then the flap was put back in place (bottom).

The modem 1962

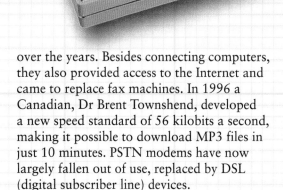

Acoustically coupled modem
Devices like this one, which first appeared in the early 1980s, used standard telephone lines. The transmission rate was 300 bits per second, 200 times slower than the norm today.

Modems are used to transfer information between computers via telephone lines. The first steps toward their development were taken in the 1950s, when staff in US military laboratories, working in conditions of intense secrecy, created, tested and used a device that enabled them to communicate with one another via a private network.

In 1958 Bell Laboratories built on this work to produce the Dataphone, which used the telephone network to transmit data. The project remained at the experimental stage, but opened the way to PSTN (Public Switched Telephone Network) modems. In 1962 AT&T put the first commercial modem on the market – the Bell 103, which transferred information at a rate of 300 bits per second. At that speed it would take 24 hours to download a modern MP3 file.

Multiple connections

Modems work by transforming digital information into analogue signals that can be sent along telephone wires, then reversing the procedure to convert them back into digital form for computers to read them. The two-step process is suggested in the name, which is short for 'MOdulator–DEModulator'.

In addition to the components responsible for this conversion, modems contain circuits allowing them to connect themselves into the network. Communication protocols, harmonised by the International Telecommunication Union, guarantee the smooth circulation of data.

Available either as internal cards or external boxes, the products were steadily improved over the years. Besides connecting computers, they also provided access to the Internet and came to replace fax machines. In 1996 a Canadian, Dr Brent Townshend, developed a new speed standard of 56 kilobits a second, making it possible to download MP3 files in just 10 minutes. PSTN modems have now largely fallen out of use, replaced by DSL (digital subscriber line) devices.

DIGITAL MODEMS

A modem is regarded as digital if it transmits digital signals via a digital network. Cable and DSL modems are digital by definition, even if they use cable-television infrastructure or the public switched telephone network. The term is wrongly applied if there is no matched modulation and demodulation of the signal. Technically, digital modems are not modems at all, since there is no analogue stage.

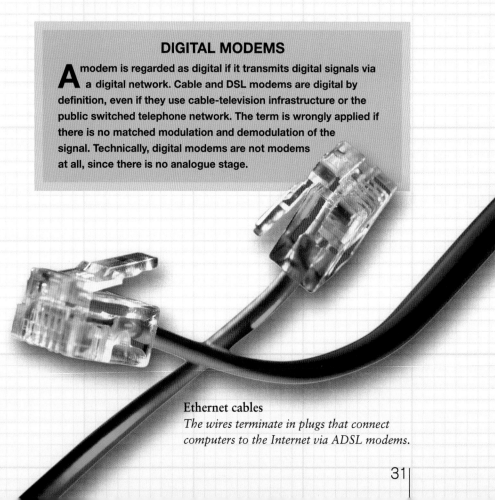

Ethernet cables
The wires terminate in plugs that connect computers to the Internet via ADSL modems.

HANG GLIDERS – 1963
Giving people wings to fly

Born of NASA experiments, hang gliders went on to conquer the world before they were rivalled by an even lighter competitor: the paraglider.

At the start of the 1960s there was a vogue for waterskiing and increasing numbers of people experimented with soaring above the water attached to kites. Many came to grief, leading an Australian named John Dickenson to seek a better way. In 1963, taking his inspiration from a picture of a Rogallo flexible wing that he had seen in a magazine, he devised the Ski Wing by fixing a similar-shaped wing to a triangular arrangement of metal tubes. The pilot hung beneath the triangle, controlling the craft by shifting his weight on the crossbar in relation to the wing. Most of the elements of the hang glider were present in this arrangement, so can Dickenson legitimately claim to be its inventor?

Disputed paternity

The mere idea of hanging under a wing and running down a slope into the wind to achieve lift-off was not new – it went back at least as far as Otto Lilienthal, a pioneer of free flight who fell to his death in 1896 doing just that. But hang gliders proper combine this notion, as Dickenson did, with the Rogallo wing principle, which officially saw the light of day on 23 November, 1948, when Francis Rogallo was granted a patent on a fabric wing that ballooned out to form two half-cones when caught by the wind. Its arrowhead shape was inspired by the delta-wing jet aircraft that were setting speed records at the time. The first model was completely flexible, but Rogallo later developed a semi-rigid model with a framework of metal tubing. This cheap aerofoil, shaped for speed, was intended to be put to many uses, serving as a kite, a parachute, or to carry personnel or supplies.

In 1961 a Californian, Barry Hill Palmer, flew suspended from a Rogallo wing, achieving lift-off after running for a few paces. Other brave adventurers queued up to attempt the feat, using more or less reliable equipment and rudimentary or non-existent controls.

Free-flight pioneer
Under his children's watchful eyes, Francis Rogallo prepares to launch a kite of his own making in Central Park, New York, in 1950. He and his wife Gertrude tested their devices on the sands at Kitty Hawk, North Carolina, where the Wright brothers had made the first powered flight in 1903.

FROM KITES TO FLEXIBLE WINGS

A passionate kite enthusiast since his childhood, Francis Rogallo (1912–2009) became an aeronautics engineer working for NACA, the predecessor of NASA. Specialising in wind-tunnel work, he dreamed of designing a cheap aircraft that anyone could afford. His employers showed little interest in the scheme, so Rogallo worked on the project in his spare time. His passion was shared by his wife Gertrude, who would eventually share the patent he took out, and by his children, so the whole family joined in his experiments. Yet Rogallo had little success in marketing the kites derived from the flexible wing he developed. In 1958 his employers at what was by then NASA finally showed an interest, hoping to use the Rogallo wing to slow down Gemini space capsules on re-entry into the atmosphere. Testing began in 1962, but NASA eventually opted for classic parachutes to brake the crafts' descent to the ocean. Rogallo retired in 1970, leaving him free to concentrate on his real love of hang gliding – some sort of reward for a man who never obtained a pilot's licence.

Hang glider in action
Stretched out horizontally, the pilot is supported by a harness attached to the wing's rigid airframe. The triangular-shaped control frame enables the pilot to steer the wing.

NASA'S BIZARRE INVENTIONS

In 1961 NASA began testing the XV-8 Flexible Wing Aerial Utility Vehicle. Nicknamed the Fleep, this 'flying Jeep' was intended for military use. It consisted of a semi-rigid Rogallo wing with a motor attached, along with four wheels fixed to a metal framework (far right). The vehicle's designers hoped it would be used to deliver personnel or supplies to areas ordinary aircraft could not reach, but in fact it never got beyond the prototype stage. The same fate awaited a second NASA initiative, the Paresev (Paraglider Research Vehicle), which was tested from 1962 to 1964. The idea behind these craft was to find a way of controlling Rogallo wings so that they could be used to recover Gemini capsules on re-entry from space (inset). Designed by Charles Richards from Rogallo's original idea, the Paresevs introduced the notion of a rigid frame. Press photographs of the devices were in time to prove an inspiration for designers of hang gliders.

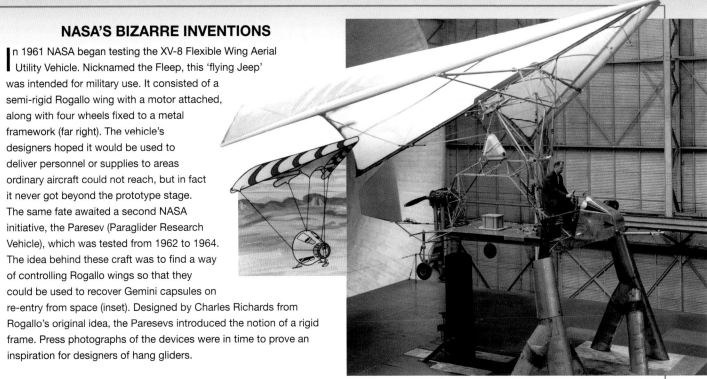

It was another Australian, Bill Moyes, who finally set out rules for the activity. He initially achieved lift-off on a No.5 Model Dickenson wing pulled behind a motorboat. With his friend Bill Bennett he soon learned free-flying techniques, launching himself first by skiing down snow-covered slopes and then on foot, facing into the wind, from cliff-tops. The duo achieved widespread publicity on 4 July, 1969, when they celebrated Independence Day in the USA by flying over the Statue of Liberty.

Overcoming risks

By the start of the 1970s free-flight hang gliding was becoming quite common. To lessen the risk of accidents, a central pole and cables

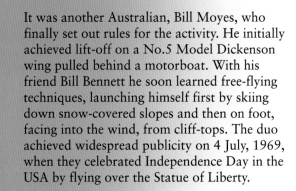

On the wing
Microlights are easier to fly than conventional aircraft but can only operate at low altitude, so are mainly used for surveillance, photography or recreation. Here (left) a pilot is escorting a flock of whooping cranes, an endangered species, on annual migration from a national park in Wisconsin to the west coast of Florida.

MICROLIGHTS

Several types of microlight aircraft are now available. The simplest are hang gliders and paragliders with motors attached, in the first case suspended in a casing under the wing and in the latter attached to the pilot's back. The first prototypes appeared in about 1975, when some enthusiasts began using lawnmower motors equipped with propellers to achieve lift-off. A number of gyrocopters and ultralight helicopters also fall within the definition of the term. The Mosquito, the first 'skytrike', was marketed from 1990 on. Designed for recreational use, microlights have become working vehicles, used by farmers for crop spraying and by photographers in search of aerial shots.

HANG GLIDERS – 1963

PARAGLIDING

In recent years hang gliding has given ground to another form of free-flight aviation. Paragliding developed out of parachuting. Parachutists had always sought to control the direction of their descent, often with little success. That began to change after 1963, when an American, Domina Jalbert, patented the parafoil, a rectangular, ribbed aerofoil inflated by the airflow entering forward-facing vents that swelled the fabric so that it started to function like an aircraft wing. The pilot, seated on a suspended saddle, controlled the craft by pulling on lines attached to the wing, or 'sail'. Meanwhile, NASA was studying ways of slowing and steering spacecraft returning to the Earth's atmosphere. Dave Barish developed the Sail Wing along the same lines as the parafoil. In 1965 tests carried out in the Catskill Mountains gave him the idea for an activity he called 'slope soaring'.

The first wing incorporating a cellular design was put on the market in 1971 by Steve Snyder as the P-1 Paraplane. Parachutists used it to train for precision landings. Paragliding enthusiasts split from traditional parachuting clubs, because their main interest was in the flying, not in descent and landing. The sport of paragliding was launched in 1978 when three friends from the Haute Savoie in France, inspired by a magazine description of slope soaring, used a ram-air parafoil to glide in free flight for a kilometre. The first sail specifically designed for free flight went on sale in 1985 and found a market among mountaineers seeking a quick descent from summits. Paragliders were cheaper and easier to use than hang gliders, and were more transportable: lacking any rigid components, they could be folded away in a bag and despite their slower speed and limited range they soon proved more popular. The first world championships were held in 1989 at Kössen in Austria.

Bird's-eye view
A paraglider soars above the Alps (left). One recent development is the sport of speed riding, which combines paragliding with skiing.

were incorporated in the design of the craft to prevent the wing from collapsing in on itself. Even so, the materials were fragile, the wings sometimes unstable and the pilots were largely untrained. There were a number of fatalities, giving hang gliding a bad reputation that nevertheless failed to discourage others. International competitions were organised, and through the 1970s and 80s rivalry built up to see who could fly the furthest.

Then habits changed. The hang gliders themselves became more reliable, and beginner pilots started taking lessons from professionals. Gradually hang gliding stopped being quite so perilous. The contests reflected the change, and simply remaining airborne ceased to be the main objective. Today pilots compete over set courses, sometimes hundreds of miles long. Some even attempt aerobatics.

The equipment used also improved rapidly. In 1978 metal reinforcement was added to the wing fabric and, more importantly, the position of the pilot was changed so that people flew the craft from a horizontal position. Then came wings shaped to maximise aerodynamic lift, as on aeroplanes. New materials such as carbon fibre were employed, and the central pole and cables disappeared. Today the most effective hang gliders have rigid wings, seemingly having lost touch with their flexible origins.

Long-distance specialist
Gil Souviron of France holds the European distance record with a flight of 540km (335 miles). Here, he is piloting a fixed-wing ATOS-VR hang glider. Employing Dacron fabric and a carbon-fibre structure, these craft are among the most efficient on the market.

ROBOTICS
Machines to serve mankind

A world away from the terrifying, sometimes humorous, robot figures in science-fiction books and films, robots in real life are machines that perform the unpleasant or boring tasks that humans cannot or do not want to do. Growing ever smarter, they will soon be able to interact with people.

ORIGIN OF THE WORD
The Czech author Karel Capek coined the word 'robot' in 1921, in a work entitled *R.U.R. (Rossum's Universal Robots)* that featured artificial humanoids manufactured to work in factories. He derived the term from the Czech word *robota*, meaning 'forced labour'.

Humanoid hero
A Westinghouse employee named Roy Wensley invented Herbert Televox, a remote-control system in a cartoon human frame. Here an executive's wife gives Herbert a hug for the camera.

At the General Motors factory in Trenton, New Jersey, in 1961, a worker unlike any other took burning hot die castings from a mould and sorted them, feeling no pain and operating without breaks and without errors. Unimate, as it was called, was a two-tonne mechanical arm equipped with pincers that carried out programmed tasks stored in its drum memory. Unimate was the first industrial robot to enter service and has had innumerable successors since in the world's industrial plants, above all in the motor industry.

Definitions of exactly what constitutes a robot vary, but all agree on certain essential traits: robots are programmable, mobile machines capable of carrying out physical tasks and sometimes reacting to their surroundings. They have articulated bodies; 'senses' – more accurately, sensors – to provide information about their environment or their own physical position; a 'brain' in the form of an implanted program; and 'hands' represented by pincers or other tools. The computer controlling them can, if necessary, be reprogrammed to make them perform different tasks. Above all, robots do work that would otherwise fall to humans – the very name implies as much.

The first steps
Robots are descended from automats, especially machines like Joseph Jacquard's mechanical loom of 1801, which could produce different patterns thanks to a

Factory hand
This Unimate 2000B model from the late 1970s was one of a family of robots mostly used to handle components on car-body assembly lines.

36

ROBOTS ON SCREEN

Since the silent-screen days of Fritz Lang's *Metropolis* (1926), robots have made memorable appearances on film. In Robert Wise's *The Day the Earth Stood Still* (1951) a flying saucer disgorges an extraterrestrial being accompanied by a faceless humanoid going by the name of Gort (*right*). *Forbidden Planet* (1956) introduced Robby, long the archetypal screen robot, who subsequently featured in many other films and TV shows. In 1977 *Star Wars* had two robot protagonists: the canny R2-D2, resembling a vacuum cleaner, and the android C-3PO.

programming system involving punched cards and a revolving drum and hooks that were raised or lowered as the cards dictated.

At the start of the 20th century, the spread of industrialisation had brought machines into an increasingly large part of the physical and mental world of humans. This was the time when the notion of the anthropoid robot endowed with consciousness – often hostile to humankind – provided the theme of many books and films.

In 1927 Roy J Wensley, an engineer with the US Westinghouse Corporation, conceived a remote-control system that could, for example, automatically measure the level of water in a reservoir and transmit the information over telephone lines, also receiving instructions in the same way. In practice, it could respond to any one of three different signals telling it, for example, to close a sluice gate. Wensley chose to humanise his device by providing it with a wooden frame shaped in the form of a cartoon silhouette of a man, which he christened Herbert Televox. The public response was astonishing. A whole dynasty of increasingly impressive Westinghouse robots, including Willie Vocalite (1930) and Elektro (1937), starred at trade fairs and international exhibitions. They did little more than generate publicity for the Westinghouse Corporation, but the public loved them.

In 1938 Willard Pollard and Harold Roselund, working for a firm called DeVilbiss that specialised in making paint sprayers for car manufacturers, took out the first patent on an articulated mechanical arm. It was the start of a long association between robots and the motor industry.

The cybernetic era

Computers and the search for artificial intelligence received a major boost in the Second World War, when mathematicians, logicians and physicists worked together to seek ways of breaking enemy codes and to calculate trajectories. In 1947 the American mathematician Norbert Wiener published an influential book in which he spelled out the new discipline of cybernetics – the study of control and communication mechanisms within self-regulating systems – whether these took the form of machines, institutions or living beings. In the same year American neurophysiologist William Grey Walter created Elsie and Elmer, two mechanical tortoises that steered themselves toward light sources, thereby demonstrating an ability to react to their environment. The theoretical bases of robotics were being laid.

In 1954 another American, George Devol, took out a patent on a 'universal programmed machine' in the form of a multi-purpose

Getting around
William Grey Walter's mechanical tortoise was designed to be self-controlling. Walter even claimed it showed signs of recognition when placed in front of a mirror.

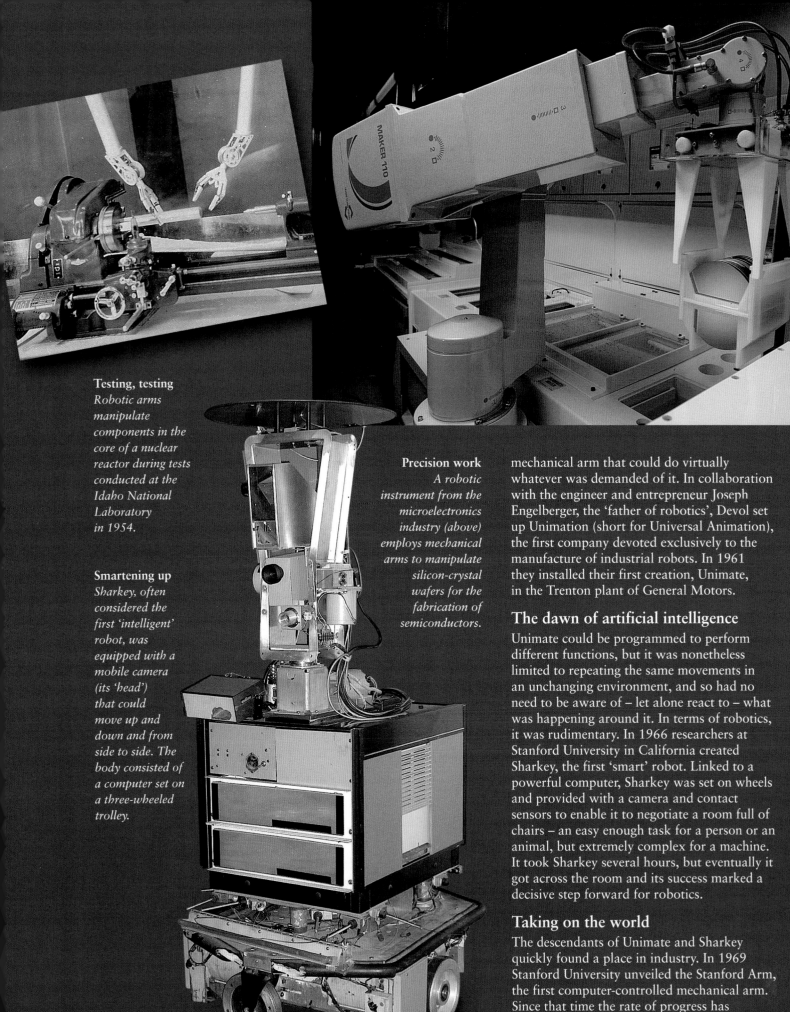

Testing, testing
Robotic arms manipulate components in the core of a nuclear reactor during tests conducted at the Idaho National Laboratory in 1954.

Smartening up
Sharkey, often considered the first 'intelligent' robot, was equipped with a mobile camera (its 'head') that could move up and down and from side to side. The body consisted of a computer set on a three-wheeled trolley.

Precision work
A robotic instrument from the microelectronics industry (above) employs mechanical arms to manipulate silicon-crystal wafers for the fabrication of semiconductors.

mechanical arm that could do virtually whatever was demanded of it. In collaboration with the engineer and entrepreneur Joseph Engelberger, the 'father of robotics', Devol set up Unimation (short for Universal Animation), the first company devoted exclusively to the manufacture of industrial robots. In 1961 they installed their first creation, Unimate, in the Trenton plant of General Motors.

The dawn of artificial intelligence

Unimate could be programmed to perform different functions, but it was nonetheless limited to repeating the same movements in an unchanging environment, and so had no need to be aware of – let alone react to – what was happening around it. In terms of robotics, it was rudimentary. In 1966 researchers at Stanford University in California created Sharkey, the first 'smart' robot. Linked to a powerful computer, Sharkey was set on wheels and provided with a camera and contact sensors to enable it to negotiate a room full of chairs – an easy enough task for a person or an animal, but extremely complex for a machine. It took Sharkey several hours, but eventually it got across the room and its success marked a decisive step forward for robotics.

Taking on the world

The descendants of Unimate and Sharkey quickly found a place in industry. In 1969 Stanford University unveiled the Stanford Arm, the first computer-controlled mechanical arm. Since that time the rate of progress has accelerated in parallel with the take-up of robots by manufacturing plants. In 1973 the German firm Kuka created Famulus, a mechanical arm with six different directions of movement. That same year, the Milacron Corporation in the USA introduced the T3 Tomorrow Tool, the first commercial robot to be controlled by a minicomputer. In 1975 Unimation responded with PUMA, the Programmable Universal Manipulation Arm. Over the following decade, the robotics industry took off, and many universities established degree courses in the discipline.

This was also the era of the space race and robots were mobilised to play their part. On 17 November, 1970, the Soviet Union landed Luna 7 on the Moon's surface. From the craft emerged a sort of bathtub on wheels, stuffed with measuring instruments. This was Lunboikodh 1, the first robot – more accurately, the first remotely controlled device – in space. In 1976 the US Viking 1 and 2 probes arrived on Mars after roughly a year's journey time, each equipped with arms that scooped soil samples from the planet's surface. Today, too, the International Space Station is famous for its manipulative mechanical arm.

Space tool
Canadian astronaut Chris Hadfield installed Canadarm2, a state-of-the-art robotic arm, on the International Space Station in April 2001. Its predecessor, the original Canadarm, had entered service 20 years earlier.

ROBOTS IN SPACE

Much of what we know about the planet Mars comes from a succession of robot explorers. The Viking probes of 1976 (below) used mechanical arms. On 5 July, 1997, the Pathfinder mission deposited a small six-wheeled vehicle called *Sojourner* which stayed in contact with Earth for a couple of months and remains on Mars to this day. In January 2004 two much larger rovers were set down on the Martian surface, both weighing some 185kg. For more than five years they have methodically explored the planet, sending precious information back to Earth.

POWERING UP

With a few exceptions – for example, when they are moved by hydraulic jacks to lift very heavy loads – robots rely on electric motors as power sources. Mechanical arms and other fixed devices are usually plugged into the mains, but mobile robots require batteries that have a limited life. Space robots use solar panels to generate the electricity they need. NASA is now exploring the concept of nuclear-powered robots.

ROBOTICS

Tireless workers

Despite their cost and complexity, robots were a rapid success story, particularly in the motor industry. They turned out to be ideal workers, able to toil 24 hours out of 24 without tiring or getting bored. Stronger, faster and more reliable than human counterparts, they could cope with extreme conditions – for example, in the toxic atmosphere of paint-rooms – and could be entrusted with the most repetitive, disagreeable and dangerous jobs.

Starting with relatively simple tasks like shifting components for packaging or feeding them into machines, robots went on to more complex jobs including painting bodywork, soldering – at which they proved more adept than humans – and the particularly intricate business of vehicle assembly.

Robots also proved to be essential servants in the field of electronics, which could never have developed as rapidly as it has without the precise dexterity of the robots employed in the production of microprocessors. They are also put to use today in some surgical procedures, where the precision of their movements is viewed as a guarantee of safety, and in biological engineering, operating in completely sterile environments and handling dangerous substances.

Automated assembly line
Robots are used in the motor industry to do the most repetitive and intricate jobs, notably paintwork and welding. The assembly line above is in a Chinese factory operated by the South Korean firm Hyundai.

THE ROBOCUP

Every year, teams of robots of different types – from devices on wheels to little anthropoid bipeds like these Sony Aibos (left) – face up to each other in an international football contest. No task is more difficult for robots than rapidly traversing a space encumbered with other moving objects, particularly when it also involves distinguishing teammates from opponents and working together with them to achieve a common objective. So the RoboCup, first staged in 1997, is seen as a good way of testing and expanding the limits of robots. Its promoters hope that by 2050 robots will be capable of beating the world's top human football teams.

ROBOTICS

> ### ROBOT PETS
> An unfamiliar animal made its first appearance in 1998 in the shape of Furby, a small, fluffy creature that was actually an automaton capable of imitating speech. The next year Sony introduced Aibo, a mechanical dog that soon became famous across the globe. Like Papero, which debuted in 2001, these robot pets require regular care and attention from their owners; otherwise, they forget everything they have been taught.

the debris left by natural disasters or fires. In 1983 a reconnaissance robot penetrated the radioactive core of the Three Mile Island nuclear plant, which suffered partial meltdown in 1979. Military robots play a vital role on the battlefield, detecting mines, putting out fires and going on scouting missions.

Robots have now been put to use almost everywhere – from the sea floor and the surface of Mars to operating theatres – and can perform almost any kind of activity. Yet they still have a long way to go. Even in assembly-line work, they cannot yet match the suppleness and versatility of the human hand, which has 25 separate fields of movement as well as all the advantages conveyed by touch. Most significantly, though, in the near future robots are expected to move out of factories and high-risk locations to play a growing part in people's daily lives. They will find a place in the complex and ever-changing environment of the home, where robot vacuum cleaners like the Roomba, introduced in 2002, are already showing the way. And they will learn to interact with us, an aspect in which robot pets are already setting the pace (see box above).

On the warpath
Droids marching out for battle in The Phantom Menace *(1999), chronologically the first of the films in the* Star Wars *saga.*

Beyond factory walls

In other fields robots are put to use wherever humans cannot go, whether because of extreme temperatures or pressures, or simply because space is too constricted – for instance, to inspect pipes. Even the most hostile environments, such as outer space or the depths of the ocean, cause robots no qualms, and they can also be used to probe

INVENTIONS THAT CHANGED OUR LIVES

Shape memory alloys 1963

Smart textiles
Researchers are exploring the possibility of creating fabrics able to respond to environmental stimuli – jackets that roll up their own sleeves in the heat, for example.

It was by pure chance that, in 1932, the Swedish engineer Arne Ölander discovered the astonishing properties of an alloy mixture of cadmium and gold. The alloy proved capable of 'memorising' the shape it had when it was heated – the form would be lost in the cooling process, but regained when the alloy was once more exposed to higher temperatures.

Over the next 30 years other alloys with similar properties were discovered, but unfortunately they all involved costly metals, such as gold, platinum and iridium, or, like cadmium, were toxic. So scientists had difficulty finding cost-effective applications for the alloys. Then an American engineer named William Buehler, who was working for the US Naval Ordnance Laboratory, happened upon a non-toxic, relatively inexpensive mix. In 1963 he took out a patent on an alloy he called Nitinol, composed of nickel and titanium (the '-nol' in the name was for his workplace).

Metal muscles

Today, shape memory alloys have a wide variety of applications in many industrial fields, including aerospace, electrical engineering, medicine, transport, construction and spectacle-making. They are used to make safety valves for deep fryers, which open and close to prevent overheating and the risk of oil fires, and self-tightening rings that expand to allow antennae to pass through them before shrinking once more as they cool.

In the 1970s orthodontists began using superelastic Nitinol wires that gradually regained their original form to reposition teeth. Since 1990 new smart alloys have come into use. Some respond to magnetic fields, others to light or electricity, opening the way for the manufacture of smart photosensors and piezoelectric micromanipulators.

FROM HOT TO COLD

When heated, the atoms composing shape memory alloys move slightly apart from one another without breaking the chemical bonds that connect them. The crystals that make them up distort, causing movement within the structure. So long as they do not melt, the materials can repeatedly shift shape, taking one form when heated and another when cooled.

Art in motion
A sculpture made from a shape memory alloy responds to being warmed by regaining its original shape. Some alloys can 'remember' two separate states, reverting to one or the other as the temperature rises or falls.

Mechanical aid
A musician uses a synthesiser to tune his drums (left). The instruments are linked to microphones, which transform the acoustic vibrations into electrical impulses that the synthesiser turns back into sounds.

The synthesiser 1964

In 1963 Robert Moog met the composer Herbert Deutsch, who was searching for new sounds to enrich his musical palette. At the time the American engineer was working on the theremin, an electronic instrument invented in 1919. Born of Deutsch's ambitions and Moog's researches, the Moog Modular Synthesiser made its first public appearance at a convention of the Audio Engineering Society in 1964.

The principle behind the instrument was a simple one. Oscillators, generators, modulators and amplifiers produced predetermined sound waves. Using a keyboard to control the effects, musicians could, in theory at least, obtain an infinite range of sounds by combining them with one another and then using filters to suppress some of the elements involved.

The invention sparked interest from the start, soon magnified by the success of Walter Carlos's 1968 album *Switched-On Bach*, entirely performed on Moog synthesisers.

Over the following years, 'Moogs' became standard features of pop as well as experimental music. In late 1970 the introduction of transportable Minimoogs brought the synthesiser, which had previously only been used for recording, out of the studios and onto the stage for live performances.

Endless possibilities

Further technical advances followed. In 1973 John Chowning invented the technique of frequency modulation synthesis, using only two oscillators compared with the multitude previously employed. Other players experimented with adding, rather than subtracting, layers of sound by combining sound waves instead of filtering them out.

By the early 1980s the adoption of a common music-industry protocol made it possible to connect computers to electronic instruments, opening the way to a whole gamut of new possibilities that musicians have only just begun to explore fully today. Many well-known artists, from the French musician Jean-Michel Jarre to the English rock band Pink Floyd, have made extensive use of synthesisers in their work.

Kraftwerk in concert
Seen here in a live performance in Bern, Switzerland, in 2004, the German group Kraftwerk used Moog synthesisers for much of their album Autobahn, an international hit when it came out in 1974.

SAMPLING

In the 1960s the Beatles, the Rolling Stones and others experimented with the Mellotron, an electronic instrument that used pre-recorded tape samplers but was cumbersome to use. The advent of digital sound in the 1980s facilitated the widespread use of sampling, which consists of reproducing synthetic or instrumental sounds, often featuring music loops.

HAROUN TAZIEFF – 1914 TO 1998
The man who listened to volcanoes

A pioneering environmentalist who became a government minister in his adopted land of France, Haroun Tazieff was born in Warsaw to a Polish mother and a Russian Tatar father. His forename was borrowed from the medieval Abbasid caliph Haroun al-Rashid, famous from the tales of the *Arabian Nights*. A boxer, rugby player, mountaineer, pot-holer, story-teller, film-maker and anti-nuclear protester as well as a scientist and politician, Tazieff was above else a trailblazer in the field of volcanology.

Craters of fire
An image from Haroun Tazieff's final film, Feu de la Terre *('Fire from the Earth'), released in 1993, shows smoke billowing from the crater of an Italian volcano.*

Tazieff grew up in Belgium, where in 1938 he obtained a degree in agronomy at the Agricultural University of Gembloux. He fought with the Resistance during the Second World War while studying for a second degree, this time in geology, at the University of Liège.

He discovered his lifetime passion for volcanoes almost by chance. In 1948, while working as an engineer in the Katanga tin mines in what is now the Democratic Republic of Congo, he happened to witness an eruption of Mount Kituro. The experience changed his life. From that time on, showing courage that sometimes bordered on the reckless, he made a habit of probing the craters of active volcanoes around the globe. He filmed lava fields and titanic explosions, seeking to elucidate the mechanisms that caused volcanic phenomena, which he viewed as release valves for the Earth's huge store of thermal energy. Above all, he sought to predict forthcoming eruptions in order to save lives.

Feet on the ground

In 1955 Tazieff met the painter Pierre Bichet, who thereafter became the photographer and cameraman on all of Tazieff's expeditions. Together they shot a documentary, *Les Eaux souterraines (Groundwater)*, that proved widely successful on its release in 1957.

The following year Tazieff led an official expedition to Mount Nyiragongo, also in the Democratic Republic of Congo. That same year, 1958, he travelled to Faial, a volcanic island in the Azores group, then to Chile in 1961 and to Costa Rica in 1964, where he

PUTTING IDEAS INTO PRACTICE

In 1981 France's President François Mitterrand appointed Tazieff commissioner for the study and prevention of natural disasters. Three years later, still charged with the task of averting natural and technological catastrophes, he was made Secretary of State, holding the position until 1986. Two years later the French Minister of the Environment commissioned him to prepare a report on risks specifically affecting the mountainous Isère region in the Alps of eastern France. His passion for politics, combined with his environmental concerns, led him to play a part in the foundation in 1990 of a new 'green' party in France called *Génération Écologie* ('Ecology Generation').

Lava fountain
The eruption of Sicily's Mount Etna in July and August 2001 lasted for 24 days and was marked by extensive lava flows, the formation of dense clouds of smoke and the emission of huge plumes of ash.

helped to avert a potential catastrophe by recommending protective engineering works at the foot of Mount Irazú. He was in Italy in 1970, the Comoro Islands and Reunion in 1972, Iceland in 1973. In Guadeloupe in 1976, he insisted that there was zero likelihood of an eruption of Mount Soufrière, but was overruled by the island's governor who insisted on evacuating thousands of people – needlessly, as it turned out.

A talented populariser

Tazieff took his ideas to a mass audience through books such as *Craters of Fire*, *Caves of Adventure* and *Nyiragongo: The Forbidden Volcano*, as well as a series of films that showed him walking on hot ashes or perched on the edge of chasms. But above all he helped to establish volcanology, which was still in its infancy in the 1960s, as a serious science, playing a part in the establishment of the International Institute of Volcanology in Catania, Sicily.

In 1972 Tazieff was offered the post of director of research at France's National Centre of Scientific Research, and later a special volcanology department was created for him at the Physics Institute of Paris. He became a naturalised French citizen in 1971, having settled in Paris in 1960. By the time of his death on 2 February, 1998, he was generally acknowledged as the world's leading authority on volcanoes.

THE DIFFICULT ART OF FORECASTING

In the year 2010, half a billion people around the world were living in the shadow of active volcanoes, almost all located on the edges of the major continental plates making up the Earth's crust. Scientists have identified some 1,500 separate sites where volcanic activity has been observed in the course of the last 10,000 years. In recent times researchers have developed a range of instruments designed to give advance warning of trouble, including seismometers that measure, among other things, low-frequency signals that can precede eruptions, spectrometers that analyse the composition of volcanic gases, extensometers that examine subtle movements in the form of the mountains, and satellite images that track changes over lengthy periods of time. Even so, the present state of knowledge does not allow scientists to work out exactly when an eruption will take place.

On-site observer
A researcher studies Mount Etna at first hand during the 2001 eruption. Direct observation is still an important tool needed to complement the information obtained from sensors, seismographs and video cameras.

CARBON FIBRE – 1964
Spinning fabrics from carbon filaments

In the 1960s a new entry was added to the long list of man-made materials. Carbon fibre gave fresh momentum to the exploration of synthetics and radically changed the aeronautics industry.

Wrapped up
Thousands of individual carbon fibres are twisted together to form a thread that can be wound onto a spool. The sample (top right) comes from one of the very first batches produced by Courtaulds in 1966.

In 1909 the first plastics, such as Bakelite, were introduced. In 1913 steel, alloyed with chrome, became stainless. A succession of new products emerged thereafter – Duralumin in 1913, asbestos in 1916, Neoprene in 1930, glass fibre in 1931, Teflon in 1938 – enriching the range of synthetic materials inaugurated five millennia ago when copper and tin were first combined to make bronze.

Continuously searching for tougher, more heat-resistant matter, the US firm Union Carbide decided to re-examine Thomas Edison's work on the carbonisation of bamboo and cotton fibres (see box below), which burned for many hours before disintegrating. Roger Bacon, one of the firm's leading chemists, heated filaments of various materials to temperatures upwards of 500°C (932°F) in order to expel the oxygen, hydrogen and nitrogen, leaving a residue of long chains of carbon atoms.

The breakthrough came in 1957 with rayon – regenerated cellulose used in the textile industry – when experiments yielded carbon fibres strong enough to be spun into a yarn. As Bacon reported: 'They were up to an inch long, and they had amazing properties. They were only a tenth of the diameter of a human hair, but you could bend them and kink them and they weren't brittle. They were long filaments of perfect graphite.' Even so, the end product contained only about 20 per cent carbon and lacked sufficient strength to be commercially viable.

A trio of discoveries

Research continued not just in the USA but also in Japan, France and Britain. Each centre succeeded in producing high-performance fibres, all at roughly the same time. In 1964 William Watt, working at the Royal Aircraft Establishment at Farnborough, Hampshire, used a new precursor (as the raw materials involved are known) – polyacrylonitrile (PAN), an organic polymer. Production quickly got under way. In the same year Akio Shindo of Japan's Governmental Industrial Research Institute in Osaka used the same precursor to synthesise fibres three times stronger than those obtained from rayon. At Union Carbide, Roger Bacon and Wesley Schalamon developed a new process that involved stretching the fibres at very high temperatures over 2,800°C (5,000°F), enabling them to orient the graphite layers to the fibre's axis.

Unparalleled strength

Manufacturing processes quickly improved, producing high-performance fibres mostly made from PAN or from pitch derived from coal tar. These were used to make materials that had 15 times more tensile strength than steel wire while weighing 75 per cent less.

A BAMBOO PRECURSOR

One day in 1878, Thomas Edison went fishing in Wyoming with some friends. When one of them accidentally dropped his bamboo rod in the camp-fire, Edison noticed that the fibres did not burn up at once, but instead gave off a brilliant glow before they ultimately blackened. The sight suggested to him that the incandescent carbon filaments involved might be suitable for use inside light bulbs. With that idea in mind, he founded the Edison Electric Light Company, which would later become General Electric, taking out a patent on his bulb in 1879. He subsequently tested other substances to find out which would provide the most reliable carbon filaments. Cotton was widely used until the 1910s, when it was largely replaced by tungsten.

CARBON FIBRE – 1964

REINFORCED POLYMERS

Coated with a layer of plastic, carbon fibres – with the atoms aligned along their long axis – provide a solid base for composite materials. The plastics combined with them were at first mostly epoxy resins, developed in the 1930s, which produced materials that were light in weight and robust; their main drawbacks were their high cost and a tendency to turn yellow when exposed to ultraviolet rays. Polyester, which was introduced in the 1950s, is cheaper and has now become the most commonly used substance for manufacturing the products.

Carbon fibres, then, are 15 times stronger than the finest steel alloys: in strength-to-weight ratio, they are the toughest man-made material ever developed. They do not expand or rust or lose their shape, and they can be rolled up and woven. They can be combined with other ingredients, notably epoxy resins or polyester, to form powerfully resistant composite materials.

ON THE MOON

The spacesuits worn in 1969 by the Apollo 11 astronauts to walk on the Moon were made from glass fibre, Teflon and high-performance carbon fibre.

Reinforced material
Carbon fibre threads – seen above under a microscope – are combined with resins to create composite materials that have a very high strength-to-weight ratio. Britain's Chris Boardman (right) rode a Lotus Type 110 carbon-composite bicycle in 1996 to set a world performance record of 56.375km (35.030 miles) for the longest distance cycled in an hour.

CARBON FIBRE – 1964

CARBON FIBRES IN THE SKIES

By 2005, some 3,000 tonnes of carbon fibre were used in the aeronautics industry, about 13 per cent of total global production.

Mass production of carbon fibre products got under way in the 1970s, first in Britain and then in the rest of the world. Initially the main market was the sports and leisure industry, where carbon fibre was used to make tennis and badminton racquets, tent frames and masts for sailing boats. Fishing rods, golf clubs, bicycle frames, kayaks and skis soon followed. On its own, though, carbon fibre had a major drawback: it tended to split and crack when subjected to violent shocks.

From sport to space

The aerospace industry was taking a keen interest in the remarkable tensile strength and heat resistance of the new composite materials, which also had the advantage of durability. Developments took a technological leap forward, as carbon fibre products were moulded into high-performance helicopter blades and rotors that could withstand the increasing power of engines and turbines. The fibres were coated in a carbon die, creating carbon–carbon alloys that could resist temperatures of up to 3,500°C (6,300°F), enough to equip them for use in the combustion chambers of rocket engines.

They were employed for the nose cones of space shuttles, as well as for the leading edges of the shuttle's wings and to make the heat shields that protected the craft on re-entry into Earth's atmosphere. On aircraft they served for brake pads and parts of the fuselage. Plastics reinforced with carbon fibre made it possible for increasingly big jet airliners to take to the skies. The Boeing 787 and Airbus A380 are both made of fibre-reinforced plastics, respectively in 60 per cent and 50 per cent ratios.

Multiple applications

Since the start of the 1990s, the chassis and brakes of racing cars have contained carbon fibre. Boats have benefited, becoming faster thanks to stronger masts and lighter hulls. By the later years of the decade, carbon fibre was finding its way into machine parts, textiles, printing presses, high-pressure gas reservoirs, felt insulation and windmill sails, as well as being used in the construction of bridges, shelters and earthquake-proof buildings.

Combining strength and lightness
Carbon fibre is employed in the manufacture of some golf clubs (left), particularly in the head and the tip of the shaft. It is also used in the chassis of Formula 1 racing cars, such as this Ferrari F60 with a carbon fibre and honeycomb composite monocoque.

In medicine, the material's biocompatibility has made it suitable for the manufacture of prosthetics. Because they do not show up under x-ray, carbon-fibre composites are used to make the tables in radiography departments. Today engineers are considering putting the fibres' conductive properties to use in vehicles powered by solar electricity and in energy-efficient buildings. The one disadvantage of carbon-fibre material is that it is not biodegradable, making it difficult to recycle.

Jumbo jet
The Airbus A380 (above) is currently the world's largest civil aviation aircraft, capable of carrying up to 800 passengers. To reduce weight, composite materials are used for the wings, tail surfaces and sections of the fuselage. Below: a US Space Shuttle re-enters the Earth's atmosphere protected by a reinforced carbon-carbon heat shield.

A DANGER TO HEALTH?

Asbestos first made people aware that synthetic materials can carry serious health risks, when it was shown that prolonged exposure to inhaling asbestos fibres could cause the lung disease known as asbestosis. Similarly, some very fine carbon fibres, less than 3 micrometres in diameter, could penetrate deep into the lungs if inhaled, but average thicknesses are more in the order of 7 or 8 micrometres, so the dangers to health are generally considered to be less than those posed by asbestos. Yet it is hard to be certain since problems can arise as much as 50 years after exposure. In the meantime, the results of toxicological tests have to be treated with caution and preventative measures are recommended, particularly in factories where fibres can be split into microfibrils that are easily breathed in by workers.

FRANÇOIS JACOB AND JACQUES MONOD
The intelligence of genes

Biologists had long suspected that genes were more than vectors of heredity, but they had little idea of the finesse of the processes controlled by these seemingly basic units. What Jacob and Monod revealed was the extraordinarily complex mechanisms of genes.

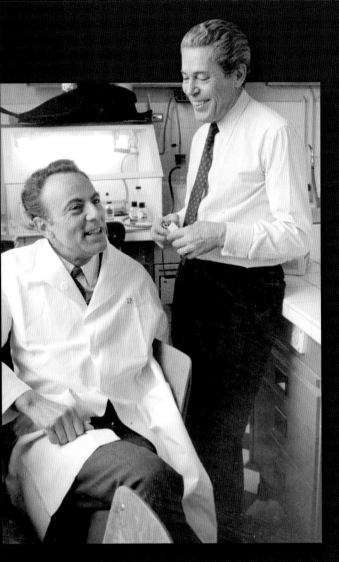

Dynamic duo
François Jacob (on the left) was born in 1920 and qualified as a doctor in 1947, joining the Pasteur Institute three years later. From 1964 until 1991 he was a professor of cellular genetics at the Collège de France in Paris. He is a member of the French Academy and the Academy of Sciences. Jacques Monod (on the right) received a doctorate in natural sciences in 1941, then spent most of his career at the Pasteur Institute. He served as its director from 1971 (the year this photograph was taken) to 1976. Monod also occupied the chair of molecular biology at the Collège de France from 1967 to 1973.

In 1947 Jacques Monod (1910-76) made an unexpected discovery: when he cultivated *E. coli* bacteria in a Petri dish containing two sugars, glucose and lactose, they absorbed the glucose but initially left the lactose untouched and stopped multiplying. Did they lack the necessary enzymes to digest it? But then, 20 minutes later, they started to break down the lactose and the culture got back on track.

This was the starting point for a line of research that Monod would follow with François Jacob, a colleague ten years his junior, under the supervision of André Lwoff, the director of microbiology at the Pasteur Institute in Paris. The work would win the three of them the Nobel prize for medicine in 1965.

A complex mechanism

The first thing Jacob and Monod discovered was that during the 20-minute delay the bacteria was producing galactosidase, the enzyme it needed to digest the lactose. The mechanism involved was worthy of a complex bureaucracy. First a precursor of the enzyme had to fix itself onto a group of genes collectively known as the lac operon. These authorised the production of the enzyme via RNA messenger molecules (mRNA). The lactose was then broken down into its two component parts, glucose and galactose. The evidence suggested that the bacteria preferred glucose; when they were offered lactose, it took time for them to assimilate it.

Oddly enough, it seemed that the take-up of the lactose began only after the glucose had already been absorbed because the *E. coli* bacteria needed to produce a specially adapted enzyme. Even more surprising, it turned out that if the lactose level started to decline before they had finished their 'meal', the lac operon could call on the services of special proteins called repressors that were activated when the bacteria had had their fill, cutting off the supply of the enzyme. The discovery of this mechanism cast light on the way that cells function, enriching understanding of the metabolism and physiology of living organisms.

The pyjama experiment

The work of the two biologists gradually widened in scope, winning widespread attention because it ran against some generally held views. Scientists had long believed that cells adapted themselves to their environment once and for all, 'knowing' what they needed to do to survive and develop. Cell behaviour, it was thought, was automatic; so, for example, the galactosidase enzyme could be expected to

---

RESISTANCE FIGHTERS

The early careers of François Jacob and Jacques Monod had marked parallels. A medical student when the Second World War broke out, Jacob travelled to London to join the Free French forces and saw four years' service in North Africa as an army doctor. Monod was a chief of staff with the Resistance inside France, playing an active role in the liberation of the country.

appear only after the glucose reserve had been used up. But the process turned out to be more complex than that: Jacob and Monod showed that cells sometimes produced galactosidase even when they could not use the sugar they were offered. It seemed now that the 'cellular bureaucracy' could change its own rules in an apparently random fashion, a phenomenon that the two scientists explored in two celebrated books, *Chance and Necessity* (1970) and *The Possible and the Actual* (1981). The bacteria absorb the glucose by an inbuilt reflex because it is their staple diet; but when it comes to the lactose, with which they are not familiar, they employ an adaptive mechanism, requiring instructions on what to do.

Following up the idea, Jacob and Monod undertook the so-called 'pyjama' experiment, grafting properties of one type of bacterium onto another. In doing so they discovered there are two sorts of genes: structural ones that code for the synthesis of proteins and enzymes, and others, composing the operons, whose function is to authorise or repress the functioning of the structural ones. Genes, it seemed, could regulate other genes: this discovery marked a significant step forward in the history of biology.

The two researchers had suggested the full extent of the finesse and complexity of the mechanisms that could be glimpsed in the double-helix structures of DNA and RNA. Their names are now inseparably linked in the roll call of biological greats: people talk of Jacob-and-Monod in the same way that they speak of Crick-and-Watson.

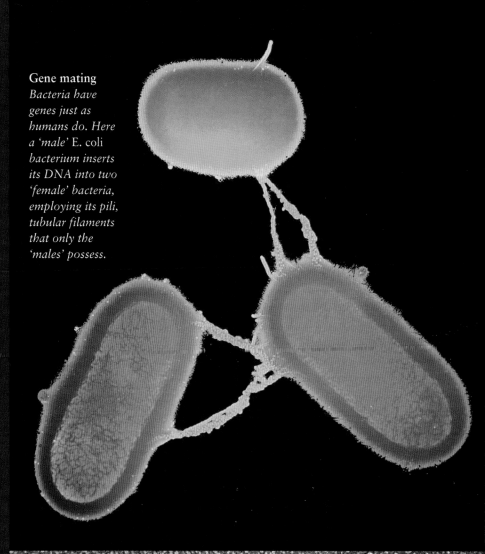

Gene mating
Bacteria have genes just as humans do. Here a 'male' E. coli bacterium inserts its DNA into two 'female' bacteria, employing its pili, tubular filaments that only the 'males' possess.

Chain of life
This fragment of messenger RNA (right) comes from a midge of the Chironomus *species.*

Prize-winners
André Lwoff flanked by Jacob and Monod on 5 December, 1965, just before flying to Stockholm to collect the Nobel prize for medicine.

WORD PROCESSING – 1964
The computer takes over from the typewriter

The very first word processing system was born in 1964 in a German subsidiary of the IT giant IBM. It was developed for internal use only to deal with an ever-growing pile of paperwork and also to produce the company's own technical brochures. Commercial word processing systems appeared in the mid 1970s. Since then many different applications have been added as word processing has become accessible to everyone.

Work tool
Students using an Apple III computer, a model launched in 1980 with more memory than its predecessor and a built-in floppy-disc drive. Technical problems, caused partly by the lack of a fan and air vents, led Apple to discontinue the model four years later.

In 1976 the American firm Wang developed computers specialising in word processing, designed for multi-user file sharing. In effect, these were typewriters provided with a central processing unit and a memory to handle text capture. Despite their high cost, the machines found a ready market because of the amount of time they saved in business. They also offered an on-screen choice of typefaces, video inversion and a wordsearch function.

Though the Wang machine was snapped up for business use, the system's mix of word, data and image processing was not right for the general public. To meet the demands of that market, the machine/user interface had to evolve away from text-editor programs, bristling with lines of code, to systems that could deliver an interactive on-screen presentation of the text.

Major players enter the market
In 1975 the Hungarian-American electrical engineer Charles Simonyi showed the way to a more user-friendly approach by developing, for the Xerox corporation, the Bravo program, the first WYSIWYG ('What You See Is What You Get') text processing system. In doing so he laid the foundations of modern word processing. AppleWriter followed in 1979, created by one-time spacecraft component designer Paul Lutus, while Seymour Rubinstein, founder of MicroPro International, created WordMaster, which developed into WordStar. Improved from version to version, the two products dominated the market until 1990, when the joint appearance of Word and Windows 3.0 signalled their decline. By that time Charles Simonyi was working for Microsoft, and from 1983 on he used the experience built up working on Bravo to develop the first version of Word for MS-DOS.

From strength to strength
The first products built around a graphic interface now appeared on the market, allowing the user to insert text at a spot marked by a cursor, accessed with the aid of a mouse. Apple led the way here, introducing MacWrite in 1984. Meanwhile Microsoft was building up a reputation for its office tools.

WORD PROCESSING – 1964

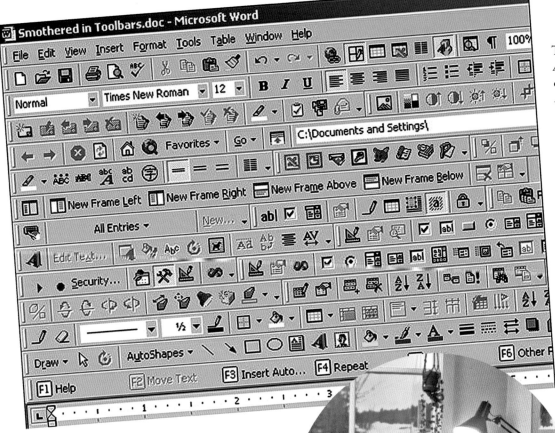

Text editing made easy
An extensive choice of toolbars (left) simplified the business of making changes in bodies of text. Soon people of all persuasions were making use of the new system, among them the radical activist David Dellinger (seen below in 1985), a representative figure of the 1960s' counter-culture in the USA.

Numerous floppy disks were needed to install these programs, but when they were up and running they brought dramatic benefits to those who used them. Dark screens filled with command lines gave way to more user-friendly interfaces controlled by the mouse. It became as easy to amend copy as to type it in. The magic of cut-and-paste changed working habits by opening up previously unthought of ways of originating and organising documents. It was goodbye to typewriter correction fluid and the personal presentation of pages of copy; in future everything would be done on screen. Typists, secretaries, employees and executives all benefited from the innovations, which increased in pace after the first PCs – personal computers – started appearing on the market.

In 1990 Microsoft took a leaf out of the Apple Mac book; Word and Excel became easier to use when they were offered up in the Windows format. At roughly the same time the firm introduced Microsoft Office, a suite of applications combining the Word, Excel and Powerpoint programs. Word processing was now compatible on both PCs and Macs, creating ever more powerful tools. New applications succeeded one another, from spellchecks to page design – so many, in fact, that few people bothered to use all of them.

THE ADVENT OF SPREADSHEETS

VisiCalc, the first spreadsheet program, came on the market in 1979. The brainchild of a US firm called Software Arts, it was designed for use on the Apple II computer. VisiCalc proved an immediate success, rivalling calculators as a way of performing calculations on rows of figures and quickly establishing itself as an ideal tool for accountants and financiers. Lotus 1-2-3 had its moment of glory in the years after 1983, and Borland's Quattro Pro arrived in 1989. The best-selling spreadsheet programme of all, Microsoft's Excel, was introduced in 1985.

INVENTIONS THAT CHANGED OUR LIVES

Acrylic paints
1963

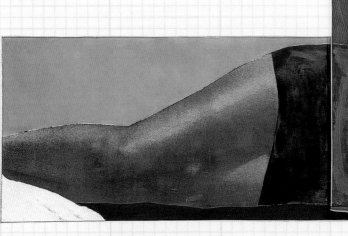

Suddenly Last Summer
The French pop artist Martial Raysse used acrylic for his 1963 work, Soudain l'été dernier, now in the collection of the National Museum of Modern Art in the Pompidou Centre in Paris.

The search for a new type of paint got underway in the 1920s, inspired by the complaints of Mexican mural painters, including Diego Rivera and other respected masters, that oil paints and traditional fresco techniques were not suitable for outdoor work, being too susceptible to the ravages of bad weather. The hunt was taken up not just in Mexico itself but also in the USA, where one of the group of artists, David Alfaro Siquieros, had a New York City studio.

Acrylics, which dry quickly and have a mild odour compared with oil paints, are made by mixing traditional pigments with an acrylic resin derived from acrylic and methacrylic acids. The first acrylics were developed in the late 1940s for house painting, but it took until 1963 for artists' versions, with a consistency similar to oil paint, to be made commercially. The first, marketed as Liquitex, was produced by the US company Permanent Pigments.

> **A GIFT TO ARTISTS**
>
> From a painter's point of view acrylic paints have two big advantages over oils: they dry very quickly, and they can be diluted with water. These qualities were to exert a considerable influence on the style of a number of artists, Mark Rothko and Andy Warhol among them.

Photochromic lenses 1964

The photochromic process allows spectacle manufacturers to produce lenses that darken in sunlight, adapting automatically as the wearer moves in and out of shade. Photochromic glasses were patented by the US company Corning in 1964, following research done by R H Dalton, W C Armistead and S D Stookey. The first lenses were made of glass, but other materials such as nylon and polycarbonates are also now used. The first pairs went on the market in 1967 under the brandname Photogray. In the 1990s Transitions photochromic lenses, made from plastics, were an international success story.

The principle involved in the process is straightforward enough. Whatever material they are made from, the lenses contain pigments that react by darkening when exposed to ultraviolet rays and lightening when the intensity of the radiation decreases. The result is easy on the eyes, protecting the retina from the harmful effects that can result from exposure to bright sunlight.

The mini-skirt 1965

London's Chelsea district gave the mini-skirt to the world. It was there that Mary Quant, a one-time student at the city's Goldsmiths College, had her boutique, Bazaar, selling clothes of her own design. After years of experimenting with steadily shrinking hemlines, in 1965 she introduced the mini-skirt, named for her favourite car, the Mini, intending it to be practical, cheap and easy to wear. The style was an immediate success; that same year the Parisian fashion designer André Courrèges produced an haute couture version. But not everyone was impressed: Christian Dior commented that 'the knee is the ugliest part of the female body'.

Liberated women

Yet there was an evident symbolism at work, for women's emancipation had long been linked to the length of their skirts. Ankles were first displayed around the time of the First World War. In the inter-war years sportswomen like the American tennis star Helen Wills Moody promoted knee-length garments, offering greater freedom of movement. The invention of tights in 1959 did away with the worry of revealing suspenders and the tops of stockings, making shorter skirts possible.

The new look was adolescent and androgynous, as personified by Mary Quant's best-known model, Twiggy. It rapidly became the uniform of 1960s young women, although not all of them carried it off quite so well as the models. Showing off the legs was a way of affirming that a girl was at ease with her body – and, by implication, was liberated. Like the lengthening hair fashionable for young men at the time, it was a challenge to established conventions, intended to shock the older generation and those of conservative sensibilities.

Since that time the mini-skirt has gone in and out of fashion, tending to return to vogue in times of economic prosperity as a symbol of optimism. Even so, the skirts are not for everyone, since they do not suit every pair of legs.

New faces of fashion
The best-known model of the mini-skirt era was Twiggy (on the right, above) here posing in a duo wearing Mary Quant dresses.

CONTROVERSIAL TO THIS DAY

Even today, mini-skirts can cause controversy. In 2008 Uganda's Minister for Ethics and Integrity called for them to be banned in case they caused traffic accidents. In 2009 in Guatemala, the Minister of the Interior barred them from his ministry in order to preserve its 'good image'. And in 2010 the mayor of the Italian seaside town of Castellammare di Stabia sought to crack down on them in the interests of 'preserving public decency'.

TIDAL POWER STATIONS – 1966
Electricity from the sea

In Brittany, upstream from the coastal resorts of Dinard and St. Malo, a busy road crosses the River Rance. The bridge looks ordinary enough at first sight, rising a few metres above the water level, but underneath the highway lies a power station unlike any other in the world: the Rance tidal power station.

Turbines at work
For people watching the power station at work from above, the chief visible signs of the turbines are the whirlpools (below) created as water is sucked into their blades (left).

The Rance power station operates on the same principle as the tidal water mills used in the Middle Ages to grind grain. It is constructed in the form of a dam 32m (105ft) high. When the tide comes in, the sea fills an upstream basin 22km² (8.5 square miles) in extent. As it goes out, the sluice gates are closed, creating a difference in the water level on the two sides of the dam. When they are reopened, water pours through, activating 24 turbines linked to alternators that each generate 10,000kW of electricity. The basin empties as the tide retreats, continuing to produce electricity. When the tide comes in again, the process is reversed and a new cycle gets under way.

An old idea

The concept dates back at least to 1897, when a civil engineer by the name of Pilla-Deflers suggested it to the provincial governor responsible for the area. He pointed out that the site was ideally suited to producing power, given the unusually big difference in water levels between high and low tide, averaging 8.2m (27ft), as well as the sheer volume of water passing through – 18,000m³ a second,

THE HEART OF THE OPERATION

The electricity generated by the Rance station is produced by 24 Kaplan horizontal turbines linked to the same number of alternators. Invented in 1912 by Victor Kaplan, an Austrian engineer, these devices possess four blades whose angle of incidence can be altered while they are in operation, so they continually adapt themselves to the flow of the current. Each turbine-alternator pod is 13.85m (45ft) long and 6m (20ft) across, weighing 467 tonnes. The blades turn at a rate of 93.75 times per minute.

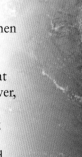

TIDAL POWER STATIONS – 1966

Kislaya Guba in Russia, among other places, but the high construction costs involved have dampened enthusiasm for the idea. Tidal power almost certainly has a future, but estuaries may not be the best place to exploit it. In Britain a barrage to harness the tidal power of the River Severn has been proposed, but remains fraught with controversy because of its potential environmental impact.

Since January 2004 an experimental plant has been operating at Kvalsund in the north of Norway, tapping the power of submarine tidal currents below the sea's surface. Other, similar projects are being tested elsewhere. Experts estimate that tides could produce up to 100GW of electricity worldwide – equivalent to 100 nuclear reactors.

Bird's eye view
The Rance tidal power station is located 4km (2.5 miles) upstream from the Atlantic coast at a point where the river is 750m wide. Looking from the foreground of the photograph, there is a lock providing passage for boats, then the power station itself in the deepest part of the river. Beyond that a dyke continues the dam, then comes a mobile barrage equipped with six floodgates. Over all runs a dual carriageway road, which traverses the lock across two vertical-lift bridges.

equivalent to ten times the flow of the River Rhone. But the engineering challenges were considerable and the project was sidelined for decades. In March 1957 it was finally declared a matter of public utility and the task of building and running the station was assigned to Electricité de France (EDF), at the time a state-run enterprise. Work started in 1960 and the station entered service six years later.

Tapping tidal power

Fifty years on, the station produces an average of 600 gigawatt hours (GWh) of electricity a year – a mere drop in the ocean compared with the total amount of hydropower generated by dams in France, which in 2008 reached 65,000 GWh. Its output is also dwarfed by the 24,000 GWh generated by the larger French nuclear power stations. On the plus side, the Rance power station taps an energy source that is endlessly sustainable and non-polluting, although it has had some negative environmental effects in disturbing the estuary's ecosystem through silting.

Elsewhere in the world, experimental tidal power stations have been built at Annapolis Royal in Canada, Jiangxia in China and

A HERCULEAN TASK

A task force of 600 workers built the Rance tidal power station for a cost equivalent to 740 million euros (£625 million). The crux of the enterprise lay in stopping the flow of water between the river and the bay so that the construction could be done 'dry'. To drain the area, two temporary dams were built and the water between them siphoned off, allowing the construction team to erect a dyke that was hollowed out along much of its length to house the turbines. Fourteen thousand tonnes of steel went into the construction.

INVENTIONS THAT CHANGED OUR LIVES

The hovertrain 1966

Twin prototypes
In all, seven versions of the Aérotrain were built. The original 01 model (right) was tested in 1965 at Gometz-le-Châtel southwest of Paris. The I-80 (above) was the final version, running on an 18km (11 mile) test track that was intended as the first stretch of a planned line between Paris and Orléans.

At the start of the 1960s, even express trains rarely went faster than 130km/h (80mph). Following the success of his hovercraft, British inventor Christopher Cockerell turned his attention to the possibilities of a hovertrain and by 1963 had developed at Hythe in Kent a test vehicle powered by a linear induction motor. By 1965 a scale model had been built, which ran on a track about a metre high, with a central section raised in a shape like an inverted T. It was exhibited in 1966, but the project failed to develop much further for lack of funding.

In France, an engineer named Jean Bertin proposed to create a hovertrain prototype capable of reaching 350km/h (220mph). The locomotive, called the Aérotrain, had no wheels; instead, like Cockerell's version, it travelled just off the ground on a cushion of air. The proposal was greeted with amazement when he put it to the board of SNCF, the French state railway system. Bertin had already worked on an all-terrain hovercraft, the Terraplane, for the French army. This employed nozzles to eject gases downward, creating a cushion of high-pressure air that lifted the craft off the ground. Designed along similar lines, the Aérotrain travelled on a monorail track with the vertical section serving to steer it and prevent derailments.

End of the line
The Aérotrain made its public debut in February 1966, when television cameras captured it travelling at 200km/h (125mph). Few people doubted that it represented the future of rail travel. In 1967 France's Minister of Transport authorised construction of an experimental track near Orléans, designed to permit speeds of 250km/h (155mph). The I-80 prototype appeared in November 1969 and plans were made for a high-speed link between Paris and Cergy, northwest of the capital.

Then, in 1974, the French government abandoned Aérotrain, claiming that it used too much fuel. In fact, SNCF had instead opted for the TGV (*Train à Grande Vitesse*), which could use the existing network. In September 1975 the go-ahead for the Paris–Lyon TGV line was the final nail in Aérotrain's coffin. Today the German-designed Transrapid in Shanghai and the Japanese Maglev hovertrains both run above the track, but they are levitated by magnetic induction rather than by air.

JEAN BERTIN – AÉROTRAIN INVENTOR

A graduate of France's elite École Polytechnique, Jean Bertin (1917-75) was an engineer with a flair for innovation. The main focus of his work was air-cushion technology. As director of Bertin and Co, he developed the Terraplane for use on land, the Aérotrain to operate on rail and also a sea-going counterpart, the Naviplane. But his most familiar invention is probably the reverse-thrust mechanism that is still used today in most aircraft jet engines.

The container ship 1968

The first purpose-built container ship was the *Hakone Maru*, which entered service with a Japanese shipping company in September 1968. The concept, which was to revolutionise international commerce and do much to advance globalisation, went back a dozen years earlier to 1956, when an American entrepreneur named Malcolm McLean had set about finding a way of optimising the time taken to load shipments in port. He decided to adapt a boat to carry the container sections of his trucks, recycling for the purpose an oil tanker that he renamed the *Ideal-X*. The vessel made its first trip in April 1956, sailing from Newark to Houston.

Maritime monsters

It took another decade before the shipping industry cottoned on to the advantages that containers offered. The construction of container ships in place of the bulk carriers that had previously been employed brought huge gains in productivity, helping to rationalise maritime commerce and forcing drastic change in working methods in docks around the world. Modern vessels can carry between 500 and 3,000 container units, with the very biggest managing as many as 11,000.

In an age concerned with sustainable development, though, container ships raise major environmental concerns. Travelling at 14 knots (about 16mph), a typical container ship uses 150 tonnes of fuel a day; at 24 knots (28mph), it consumes double that amount. Between them, the world's 15 largest container ships create almost as much pollution as all the world's cars put together. Even so, 90 per cent of sea-borne cargo now travels in containers.

> **CONTAINERS**
>
> As applied to maritime transport, a container is a large metal box of standard dimensions designed to facilitate the handling and transport of cargoes of all kinds. Size and shape are dictated by the need to make the best possible use of cargo space and also to allow the containers to travel by road or rail as well as by boat. The standard dimensions were established as early as 1961 at a length of either 20ft or 40ft (6.1 or 12.2m), a width of 80ft (24.4m), and heights of 8.5 or 9.5ft (2.6 or 2.9m). The maximum gross mass of a 40ft container is over 30 tonnes.

Recycling containers
In 1996 the Hyundai Fortune (below) was badly damaged by an explosion, but 2,000 of the 5,000 containers aboard were salvaged. Recycled, they found a number of different uses in different places: in the Afghan capital of Kabul, for instance, one now houses a dress shop (above).

PULSARS – 1968
Stars that give off radio waves

When, in 1967, British scientists first detected strange radio signals emanating from the far reaches of the universe, they initially speculated that they might be coming from extraterrestrials. But the scientists soon realised that they had discovered a new type of heavenly body: pulsars.

Pulse hunting
Antony Hewish stands in front of one of the radio telescopes of the Mullard Radio Astronomy Observatory near Cambridge (right), which he and his colleague Jocelyn Bell used to discover pulsars. The drawing (top right) is an artist's impression of the radio beam emitted by a pulsar discovered in the Crab Nebula in 1968. Astronomers can momentarily record the radio waves if they happen to sweep across the Earth's line of vision.

From the 1930s on, telescopes directed skyward have added several new entries to the list of known celestial objects, notably quasars. In November 1967 two British astrophysicists, Antony Hewish and his student Jocelyn Bell, used equipment prepared by Martin Ryle for a Cambridge observatory to analyse powerful radio sources that were being detected. While doing so, Bell noticed very brief signals coming from the Vulpecula ('Fox') constellation; the signals were regular, every 1.337 seconds, at a frequency of 81.5MHz. The two briefly toyed with the notion that extraterrestrials might be responsible, jokingly christening the signals LGM-1, for 'Little Green Man'. Then, a month later, they detected similar regular radio signals from three other sources. Discounting the possibility that four separate extraterrestrial civilisations would all have chosen the same means of communicating with the Earth at the same time, they decided the signals must be coming from natural sources.

Pulsating stars
The scientists published their findings in the February 1968 issue of *Nature*, without having managed to identify the true nature of these strange radio waves. Soon a dozen or more other sources were discovered. Astrophysicists suggested that the bodies responsible for the signals must be neutron stars – super-compact cores left after the stellar explosions known as

BEACONS OF DEEP SPACE
Because of their rapid rotation, pulsars behave like giant dynamos, generating intense electrical fields. This energy is emitted in the form of radio waves that shoot out in a narrow beam from the star's magnetic poles, whose axis is not aligned with that of the star's rotation. Like a lighthouse beacon, this cone-shaped beam can sweep across the Earth's line of vision at each turn, permitting radio telescopes to record the star's pulsation.

In a spin
B1257+12, a pulsar shown here in an artist's impression, rotates in just 0.0062 seconds. The star is orbited by three planets whose gravitational interactions affect its period of rotation.

Double take
Two pulsars emitting on gamma-ray wavelengths (left) were captured by NASA's EGRET satellite-borne telescope designed to detect gamma rays in space. Geminga appears at top left in the image, above the Crab Pulsar.

supernovas. When this type of massive star reaches the end of its lifecycle, the core collapses in on itself, shrinking in size as its rotation accelerates, like skaters folding their arms to spin faster. The neutron star then becomes a pulsar, the word itself being a contraction of 'pulsating star'.

A test for general relativity

Pulsars beat all celestial records. Their diameters never exceed a few dozen kilometres, but their mass is between 1.4 and 3 times that of our Sun. Their density is hard to imagine: 1cm^3 of matter from a pulsar would in Earthly terms weigh 100 million tonnes. The speed at which they rotate is also mind-boggling: of the 1,800 pulsars now identified, the slowest completes a rotation in 4 seconds, the fastest in 10 milliseconds.

As work on pulsars continued, more and more came to light. In 1974 Russell Alan Hulse and Joseph Hooton Taylor Jr spotted the first binary pulsar, located in the Aquila constellation, which was given the name PSR B1913+16. The existence of this system, comprising a pulsar and a neutron star, provided evidence for the emission of gravitational waves predicted by the theory of relativity. This postulated that there should be a corresponding gain in energy to balance out the loss of gravitational energy as two bodies approached one another, a phenomenon that was duly observed in the form of a variation in the frequency of the radio waves' emission. The discovery won the two American physicists the 1993 Nobel prize for physics.

QUASARS

The term 'quasar' is a contraction of 'quasi-stellar radio sources'. They are generally held to be super-concentrated haloes of energy surrounding black holes at the centre of galaxies.

THE WOMAN WHO MISSED OUT

The 1974 Nobel prize for physics was awarded to Martin Ryle for his work on radio telescopy and to Antony Hewish for having detected pulsar signals. Jocelyn Bell Burnell, who had first spotted the pulses while analysing data provided by the telescope, was not included in the award.

INVENTIONS THAT CHANGED OUR LIVES

The sailboard 1968

In 1964 an American sign-painter, artist and inventor named Newman Darby had the idea of planting a mast and sail on a surfboard, connected by a universal joint that allowed it to rotate with the wind. The original design, which he tested on Trailwood Lake and the Susquehanna river near his home in Pennsylvania, had the windsurfer with his back to the sail, which could only be turned to the side, and it was not a success. Four years later his compatriot James Drake kept the mast on a universal joint, permitting it to turn in all directions, but also introduced a wishbone boom to control the sail. His friend Henry Schweitzer bought up the rights and took out a patent on the design. It was marketed through Windsurfing International, which also trademarked the word 'windsurfer'.

Ever lighter

The first boards, made of polyethylene, were relatively heavy and hard to manoeuvre, but they still proved a success. One line of improvement lay in reducing the weight of the boards thanks to the adoption of thermo-formed polystyrene. The boards were shortened and equipped with toe-clips, permitting the rider to jump waves, marking the start of funboarding. Competitions got underway with four different types of contest: speed, with record-breakers touching 90km/h (55mph), slalom, races over set courses, and wave sailing, dominated in the early years by America's Robby Naish and France's Nathalie Simon. The first world championships were held in 1973 and the discipline featured in the 1984 Olympic Games.

The introduction of battens in 1983 and a transparent monofilm six years later improved the quality of the sails. The boards became even lighter with the adoption of a sandwich structure, alternating strips of manmade materials like polystyrene and glass fibre.

KITESURFING

Invented in 1984, the kiteboard resembles a sailboard without a mast. Kitesurfers stand on a short board equipped with toe-clips, connected by a harness to a curved, inflatable kite. To steer, they use a bar held at arms' length that is attached to the kite by four lines.

Riding the waves
A windsurfer becomes airborne off the coast of Baja California in Mexico. In 2010 the windsurfing speed record stood at 91km/h (56.5mph), slightly behind kitesurfing on 93km/h (57.8mph).

The computer mouse 1968

On 9 December, 1968, Douglas Engelbart demonstrated a very special piece of equipment to a group of information-technology professionals at the Stanford Research Institute in California. The device, later described on the patent application as an 'X-Y position indicator for a display system', consisted of a wooden body housing two perpendicular discs, whose movements were tracked on a computer screen. Two potentiometers – resistors able to translate its position into electrical currents – served to measure its movements. After five years' work on the interface between people and machines, Engelbart had come up with a brilliant gadget that would soon become known as the mouse.

Optical mice

In 1979 two Swiss computer scientists, Jean-Daniel Nicoud and André Guignard, substituted a ball that sat almost flush with the bottom of the box for the two perpendicular wheels. The ball activated three rollers, each of which acted like a notched disc. This system also effectively tracked the mouse's movements, displaying them via an on-screen cursor. The device also had two, sometimes three, buttons that could be clicked to perform specific functions. All the makings of the mechanical mouse were now in place. Two years later Xerox designed the first mouse built specifically with personal computers in mind.

In 1999 ball-operated mice gave way to the more accurate optical variety. These rely on a system combining light-emitting diodes (LEDs) and image sensors with digital readers that analyse the data provided about the surfaces on which they are moving.

The most significant recent developments have been the addition of a scroll wheel and design improvements to make the instruments easier to handle. On laptop computers, mice have been replaced by trackballs with built-in spheres to move the cursor. One advantage of the trackball is that it needs no wires, relying instead on infrared sensors to relay data.

Mighty mouse
The very first mouse designed by Douglas Engelbart was made of wood and equipped with a single button. It contained two perpendicular discs, and was linked to the computer by a pair of wires.

APPLE POPULARISES THE MOUSE

Steve Jobs, the founder of Apple, was so impressed by a demonstration at the Xerox laboratories in Palo Alto that he decided to adapt the mouse for use in the firm's computers. The first model to boast one was the Lisa, which came on the market in 1983, a year before the Macintosh.

LIQUID CRYSTAL DISPLAYS – 1968
LCD technology kills off the cathode-ray tube

The inherent magic of liquid crystals combined with researchers' ingenuity led to an entirely new way of creating images on screens. By the time flat screens had replaced the bulky cathode-ray tubes of older television sets, LCD technology was part of everyday life, used in everything from computers to mobile phones.

In 1888 an Austrian botanist by the name of Friedrich Reinitzer made a curious discovery. When he heated salts of cholesteryl benzoate, they did not behave like normal crystals. Instead, when the temperature reached 149°C (300°F) they melted, turning into a cloudy liquid, then at 179°C (355°F) the fluid turned clear. Shortly after this discovery, the German physicist Otto Lehmann revealed the peculiar structure of the substance – in their opaque phase, the salts display in places a crystalline molecular structure. The discovery led him to coin the term 'liquid crystal'.

Although they attracted the attention of many researchers at the time, liquid crystals remained a laboratory curiosity for the better part of a century. Then, in the late 1960s, a team working under George Heilmeier at the RCA laboratories in the USA developed the first practical liquid crystal display (LCD) technology. After years spent working on the project in secret, RCA's executives decided that it was time to reveal the fruits of their labours and set up a press conference on 28 May, 1968, on the 30th floor of the Rockefeller Center in New York. The group of journalists who attended gave an enthusiastic response when presented with televisions with flat screens, an innovation described at the time as 'revolutionary'.

Sharper screen images
These first LCD screens exploited the properties of liquid crystals trapped between two glass panels covered in conductive film. The crystals were sandwiched between two polarising filters to which a grill of transparent conductive wires had been added. When a current was passed through the wires, it produced an electrical field that oriented the crystals so that light either did or did not pass through them. The display surface was then divided up into a multitude of opaque or luminous elements – called pixels, for 'picture elements' – that could be used to represent numbers or letters of the alphabet.

The first commercial products that marked the entry of LCD technology into the world of consumer electronics adopted this so-called 'passive matrix' approach. In 1969 Sharp put the first calculator featuring the system on the market; remarkably energy-efficient, it needed only seven liquid-crystal cells to represent the numbers from 0 to 9. In the same year the US Air Force commissioned RCA Laboratories to develop a prototype LCD instrument panel, while NASA asked them to work on a display screen. In the early 1970s Seiko produced the original LCD wristwatch.

Improved systems
By that time a burst of intensive development was under way, encouraging fresh breakthroughs. With James Fergason of Kent State University's Liquid Crystal Unit and the Central Research Laboratories of the Swiss firm Hoffmann–Laroche leading the way, the international research community developed new, more efficient arrangements. In rapid succession TN ('twisted nematic'), STN

Trailblazer
Seiko put the first quartz watch with a six-figure LCD display on the market in 1973.

A MATTER OF LIGHTING
Because of their transparency, LCD screens need special lighting to make the image decipherable to the viewer. Reflective screens lit by ambient light offer reasonable picture quality in conditions of strong natural light but not in the dark. In contrast, so-called 'transmissive' backlit screens work well in dim or indirect light but are impossible to view in bright sunlight. Transflective screens are a compromise solution, combining a reflecting polymer sheet with back-lighting that comes on in dark conditions.

('super-twisted nematic') and DSTN ('dual-scan super-twisted nematic') technologies put in an appearance, considerably improving the stability of the image but without providing satisfactory contrast. In addition to sharpening the clarity of the display, scientists sought to increase its speed. In place of the old passive-matrix model, which had been wanting in this respect because the weak electric field generated caused the particles to react only slowly, a new active-matrix system, called TFT for 'thin film transistor', became the norm.

Active-matrix technology

The new TFT system was developed at almost the same time in the late 1970s in the USA, Southeast Asia and Europe. The pixels were no longer controlled by conductive wires; instead, each one responded to a transistor acting as a switch. To this effect, a thin sheet covered with hundreds of millions of tiny transistors was interposed between the crystals and the backlight illuminating them – hence the name 'thin film transistor'. When a current passed through one of the transistors, the corresponding pixel would light up. By

Close-up views
Liquid crystals (top) photographed in polarised light through an optical microscope. In the magnified liquid crystal screen above, the small panels – each in reality only 0.45mm square – represent just a few of the many thousands of pixels that make up an image on screen.

LIQUID CRYSTAL DISPLAYS – 1968

modulating the strength of the current, the liquid crystals contained in the pixel could be made to let through more or less light, thereby creating varying shades of grey. Colour was added by the use of red, green and blue filters that altered the appearance of the pixel. Active-matrix technology was initially not very reliable because it proved difficult to turn the transistors on and off rapidly, but that problem was soon solved by the adoption of more efficient integrated circuits that were capable of constantly monitoring the supply of electric current.

New products reach the market

In 1982 Seiko Epson created the HX 20, the first laptop equipped with a black-and-white LCD screen. A year later the same firm introduced the earliest LCD colour television set, with a screen only 5cm (2 inches) wide. In 1984 the US firm Greyhawk Systems produced a high-resolution LCD video projector, and in 1989 Japan's NEC Corporation produced the first laptop with a colour liquid-crystal display.

Over the course of the 1990s, screens gradually gained in reliability, performance and size. New technologies also put in an appearance: besides TFT there was MVA ('multidomain vertical alignment') and IPS ('in-plane switching'), which among other advantages could be viewed from wider angles, not just head-on.

On display
Giant LCD screens light up a street in the heart of Tokyo in 2002, at a time when Japan and South Korea were jointly hosting the FIFA World Cup.

THE DEMISE OF THE CATHODE-RAY TUBE

According to a German market-research institute, the traditional cathode-ray television set was passing into history by late 2009. Only 200,000 sets were sold worldwide in the whole of 2008, down from 1.3 million the previous year. Competition from LCD and plasma screens has combined with the advent of digital transmission to kill the technology off.

Screen service
A meteorologist in Toulouse, France, scans an array of LCD screens fed by the EC SX8R supercomputer, which entered service in 2008.

LIQUID CRYSTAL DISPLAYS – 1968

Mini-screen
A miniature television set featuring an LCD screen produced by Seiko in 1983.

The bigger picture
Made by Sharp, the world's biggest LCD screen, measuring 108 inches (274cm) diagonally across, went on display in 2007 at the International Consumer Electronics Show in Las Vegas.

Asian firms make the running

By the end of the 20th century the production of LCD screens was accelerating with specialist factories, mostly located in Southeast Asia, becoming hotbeds of development. The following decade saw a significant drop in the price of LCD displays, slashing the cost of a 15-inch screen by more than 80 per cent to between £200 and £300 and causing a surge in demand for the products.

By that time LCD technology was being employed for every conceivable type of display, carrying increasing information loads on computers, television sets, mobile phones, calculators, watches, cameras, household appliances, car dashboards and advertising hoardings. The revolution also stretched to visual aids of all kinds, as well as to a whole generation of new appliances that could benefit from tactile screens. In the coming near future LCD displays will be put to use in flexible screens just fractions of a millimetre thick – the first prototypes are now being constructed out of plastics.

Even so, LCD faces competition as the display technology of the future. Its main rivals are plasma, SED (surface-conduction electron-emitter display), organic light-emitting diodes and polymer light-emitting diodes (PLEDs). Each of these boasts its own particular advantages, but none for the moment is obviously superior to the others. The only certainty is that in future display screens will remain resolutely flat.

STRENGTHS AND WEAKNESSES

Compact and lightweight, LCD screens take up less space and use less energy than traditional monitors, and are also easier on the eye – all significant advantages for displays of all kinds. Their weak points, on the other hand, include a restricted viewing angle and poor reproduction of shades of black; in fact, colour reproduction on LCDs in general can be unreliable. One consequence is that for screens more than 50 inches (125cm) across, LCD is not the best choice, as the quality of the images tends to deteriorate with size.

INVENTIONS THAT CHANGED OUR LIVES

Electronic eyes 1968

In 1968 the British physiologist Giles Brindley took up one of medicine's greatest challenges: to give sight to the blind. Knowing that the region at the back of the brain, the occipital cortex, controls vision, he set out to diskover if this region could be stimulated in some way. He implanted 80 electrodes in the skull of a 52-year-old woman who had lost her sight through glaucoma. The electrodes were linked to tiny receptors placed under her scalp. The arrangement did give her some limited vision, consisting of little more than phosphenes (an awareness of sensations of light), but it was cumbersome and required the use of powerful electric currents.

Over the following decade the US biophysicist William Dobelle similarly experimented with a visual prosthesis featuring brain implants. He used minicameras fitted onto spectacles to film the visual field in front of the patient and to transmit the images to a portable computer that relayed it in the form of electrical impulses to the visual cortex. Dobelle's initial results, published in the year 2000, were considered modestly successful, permitting blind volunteers to see the outlines of objects.

Work in progress
Despite the invention of systems using cameras and video microprocessors set on sunglasses to send images to mini-receptors set in the patient's retina, human experiments with electronic eyes have not yet delivered conclusive results.

Sight for the blind
Represented by the round white patch in the image below, an artificial retina bearing a microchip has been tested on a blind animal. The technique should be ready for trials on humans by 2013.

THE SEEING BRAIN

Like cameras, eyes capture light, but seeing depends as much on the brain and nervous system as it does on the eyes themselves. The information they obtain in the form of colours, contrasts and perspectives has to be transferred through the optic nerve to the occipital cortex for the analysis that makes vision possible.

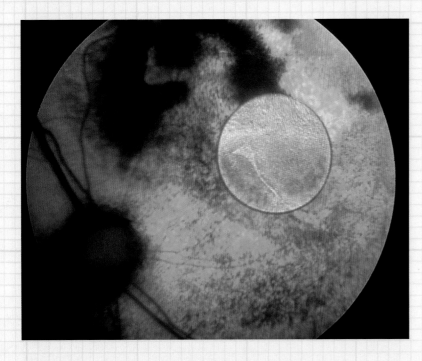

Artificial retinas
Other researchers have sought to work directly on the retina, the part of the eye where specialised cells called photoreceptors transform visual stimuli into electric impulses. In the first decade of the 21st century different prototype artificial retinas were tested on patients in Europe and the USA.

Yet for all the astonishing progress of electronics and miniaturisation, electronic eyes remain very much in the research stage. To form an image, the brain uses 100 million photoreceptors, many more than even the most sophisticated microchip. Scientists reckon that it takes 600 of the receptors in an area of about $5mm^2$ (equivalent to 25x25 pixels on a television screen) to read text, and 1,000 to recognise a face. In comparison, current experimental systems employ 64 electrodes set on the surface of the retina.

Floppy disks 1969

The first floppy disks were invented for IBM in 1969. At the time computers still relied for data feed on cards or perforated tapes, systems that had been inherited from the giant machines of the 1940s and 50s. These early arrangements could store relatively little information: it took thousands of cards or great lengths of tape to deliver instructions to perform even the most elementary operations.

The first commercial floppy disks measured 8 inches (20cm) across and were launched on the market in 1971. These new data storage tools were intended to replace the tape readers employed in the enormous early mainframe computers that were used to handle the information systems of big businesses.

The early disks consisted of a protective plastic envelope and the floppy disk itself, which was covered in a layer of iron oxide capable of storing up to 250 kilobytes of information (equivalent to about 250,000 characters), an amount that was considered remarkable at the time.

Smaller and smarter

The technology of information storage has steadily evolved over the years. In 1976 the 5.25-inch disk arrived, offering a capacity of 720 kilooctets. They quickly became essential tools and almost every computer was designed to read them. For a decade or more they were used to transfer, distribute and store data before the format shrunk again, this time to 3.5 inches with a capacity of 1.44 megaoctets. These new minidisks, introduced in 1984, also had a long lease of life despite their occasional unreliability, but they could not stand up to the competition provided by later arrivals that included zip disks (below), CDs and DVDs, and USB flash drives.

A range of sizes
Floppy disks reduced in size over the years: first came the 8-inch (top), then the 5.25-inch (centre) and finally the 3.5-inch minidisk.

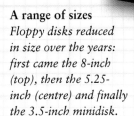

ZIP DISKS

The US company Iomega introduced zip disks (left) in 1994. Unconnected with the famous ZIP data compression file format, the disks originally offered 100MB of information storage (equivalent to 70 of the 3.5-inch floppy disks), but later 250 then 750MB versions also became available. Like floppy disks before them, zip disks have now been superseded by later systems.

QUARKS – 1969
A new way of classifying elementary particles

Since the 1950s scientists have discovered a whole universe of previously unknown particles. So how should they be classified and what are they made of? In 1969 the American physicist Murray Gell-Mann won the Nobel prize for physics for the discovery of quarks – elementary constituents of matter. Particle physics finally had a model to follow.

Elementary trinity
A proton is made up of two 'up' quarks (u) and one 'down' quark (d), held together by gluons (g).

'Three quarks for Muster Mark!' James Joyce coined the word to imitate the cry of seagulls in *Finnegan's Wake*, published in 1939, but the Irish writer could hardly have imagined that it would be adopted 24 years later for a scientific concept. Murray Gell-Mann chose 'quark' to designate hypothetical elementary particles that, associated in threes, make up the protons and neutrons of atomic nuclei.

After postulating the existence of quarks in 1964, Gell-Mann spent four years seeking to confirm their existence experimentally. His work paved the way for the drawing up of the standard model of particle physics, establishing quarks and leptons as the elementary components of matter (see box, right). The discovery of quarks enabled theorists to organise and simplify the catalogue of elementary particles, which by the late 1950s had started to look impossibly complicated.

Stranger and stranger

Since the discovery of electrons, neutrons and protons early in the 20th century, the list of particles had grown steadily longer, notably with the discovery of synchrotrons (in 1947) and other accelerators. Massive particles unknown in nature successively put in an appearance: Lambda, Sigma and Xi baryons, kaons, pi mesons and others, all of them considered elementary. In his 1961 article *The Eightfold Way: A Theory of Strong Interaction Symmetry*, Gell-Mann showed how this seemingly disparate collection could be organised through a parameter he called 'strangeness'. Each known particle was given a strangeness quotient (0, 1, –1, 2, –2, 3 or –3), calculated from its known properties, such as mass and electric charge.

Gell-Mann's achievement lay in introducing a new way of classifying the infinite diversity of matter. Classical physics sorted matter into solids, liquids and gases, basing the division on physical properties such as density and elasticity that reflected the strength of the bonds between atoms, which remain inseparable whatever the energy used to try to force them apart. Gell-Mann's system, in contrast, turned on the mysterious quarks, presumed to be the constituents of every type of particle, which could be classified according to their strangeness quotient.

Birth of the standard model

Initially there was considerable resistance to the theory. Gell-Mann (and simultaneously but separately, the Israeli physicist Yuval

Visible traces of quarks
'Top' quarks – one of six known varieties – were first detected in 1995 in the Tevatron, a particle accelerator at the Fermi Laboratory near Chicago. Produced by the collision of protons and anti-protons, top quarks disintegrate into particles whose traces can be seen in the photograph below: muons (violet), neutrinos (red) and others (green and yellow).

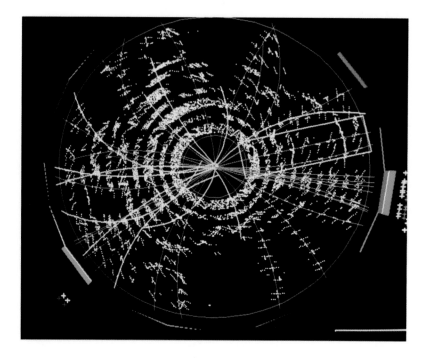

QUARKS – 1969

Long laboratory
A technician at work in the Stanford linear accelerator in California (above). The facility, which has been operational since 1966, lies 10m underground. The building housing it stretches for 2 miles (left).

Ne'eman) were able to verify experimentally the existence of 0, 1 and 2 particles, but no-one managed to detect one with a strangeness rating of 3. Then, in February 1964, researchers at the Brookhaven National Laboratory in New York discovered the baryon V-particle, which rated –3. It was a vindication of Gell-Mann's system.

The breakthrough permitted Gell-Mann to publish a complete version of his model that year, with quarks and leptons playing their part in a new table of matter. (Another US physicist, George Zweig, independently published his own model at approximately the same time.) In the wake of their work the standard model was born, to be steadily enriched and given added complexity by discoveries about the nature of matter made with the aid of ever-larger particle accelerators. Today, it is known that for every particle there is a corresponding anti-particle, with the same mass but opposing characteristics: for example, each quark is matched by an anti-quark with an opposite electric charge. Some physicists are even postulating that the known particles may also have superpartners, dubbed squarks, sphotons and so on. Others are betting on the discovery of a seventh quark, beyond the six already known.

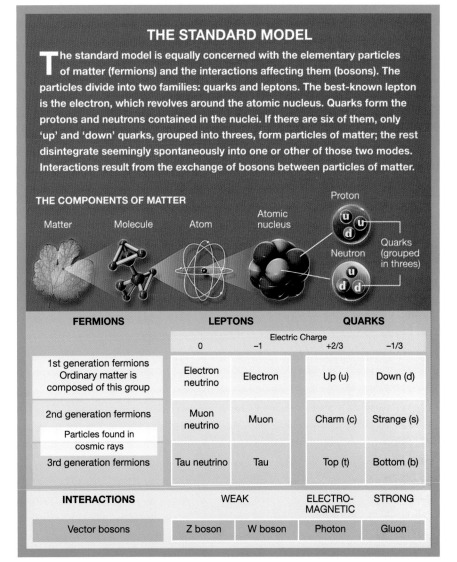

THE STANDARD MODEL

The standard model is equally concerned with the elementary particles of matter (fermions) and the interactions affecting them (bosons). The particles divide into two families: quarks and leptons. The best-known lepton is the electron, which revolves around the atomic nucleus. Quarks form the protons and neutrons contained in the nuclei. If there are six of them, only 'up' and 'down' quarks, grouped into threes, form particles of matter; the rest disintegrate seemingly spontaneously into one or other of those two modes. Interactions result from the exchange of bosons between particles of matter.

THE COMPONENTS OF MATTER

Matter → Molecule → Atom → Atomic nucleus → Proton (u u d) / Neutron (u d d) — Quarks (grouped in threes)

FERMIONS	LEPTONS		QUARKS	
	Electric Charge			
	0	–1	+2/3	–1/3
1st generation fermions Ordinary matter is composed of this group	Electron neutrino	Electron	Up (u)	Down (d)
2nd generation fermions	Muon neutrino	Muon	Charm (c)	Strange (s)
Particles found in cosmic rays 3rd generation fermions	Tau neutrino	Tau	Top (t)	Bottom (b)
INTERACTIONS	WEAK		ELECTRO-MAGNETIC	STRONG
Vector bosons	Z boson	W boson	Photon	Gluon

WERNHER VON BRAUN – 1912 TO 1977
The man who loved rockets

Germany's Wernher von Braun remains the world's best-known rocket scientist. His work on the V1 and V2 missiles led to many civilian deaths in the Second World War, but he also played a crucial part in the Apollo space mission that put men on the Moon in July 1969.

As a boy Wernher von Braun dreamed of space flight, his imagination nourished by the works of Jules Verne and H G Wells and encouraged by his mother, who gave him an astronomical telescope. The von Braun family was part of the Prussian aristocracy, with a tradition of service to the state, but no-one discouraged him from his hobby. As a student he joined the VfR, a society for people interested in rocketry, and made the acquaintance of Hermann Oberth, the father of German rocket science. Soon the group and their rockets attracted the attention of the German military, who were all the more interested because Germany was denied the use of heavy artillery by the Treaty of Versailles. After graduating from the Technical University of Berlin, von Braun was hired by the army in October 1932 to develop a liquid-fuel rocket. He quickly made his name as an exceptional engineer and organiser, first at Kummersdorf near Berlin and then at the Peenemünde test site on an island in the Baltic Sea. There, he developed the Aggregat series of rockets, labelled respectively A1, A2, A3 and A5.

Presidential visit
Wernher von Braun (on the left) with US President John F Kennedy (centre) and Vice-President Lyndon B Johnson on a visit to the Marshall Space Flight Center in 1962.

FORCED LABOUR ON THE V2s

Did von Braun know that Germany's V2 missiles were built by forced labour? By his own admission he made inspection visits to Mittelwerk, the underground factory in the Harz Mountains where the rockets were assembled by a workforce made up mostly of deportees held in the nearby Dora concentration camp. Von Braun claimed that he rarely spent more than a few hours there and had never personally seen a worker mistreated, although he did admit to having been informed from 1944 of the death of many workers as a result of the appalling inhuman conditions.

WERNHER VON BRAUN – 1912 TO 1977

Changeing allegiance
After being arrested by US soldiers of the 44th Division in June 1945, von Braun (centre, his arm in plaster) was discretely shipped to the USA under Operation Paperclip, a clandestine operation by US intelligence agencies to get hold of the finest scientists from Nazi Germany.

Servant of the Third Reich

When the Second World War broke out, von Braun proposed building a 12-tonne rocket, the A4, fuelled by ethanol and liquid oxygen. On 3 October, 1942, a prototype soared to a height of 80km (50 miles), turning Hitler's initial disdain for the project to enthusiastic backing. Von Braun's A4 became the terrifying V2 flying bomb, about 3,200 of which were eventually launched, most of them targeting London and Antwerp. In 1945 von Braun was one of a hundred or so top scientists captured by the US army and taken to the USA.

Aiming for the Moon

Von Braun spent the next four years living under cover until his presence in America was finally acknowledged in 1949. At the time of the Korean War he was ordered to develop a short-range missile derived from the V2. This became the Redstone, successfully launched for the first time in 1953.

After the USSR's success with Sputnik 1, the US authorities entrusted von Braun with the task of getting the first US satellite into space, which he achieved with Explorer 1 in 1958. The creation of NASA meant he was at last able to pursue his dream of developing a rocket to carry humans into space. The first astronauts to set foot on the Moon were carried there by the Saturn V rocket that he developed.

Von Braun was a gifted populariser. In 1952 he published *The Mars Project*, promoting the idea of a manned mission to the Red Planet. In 1970 he was appointed NASA's Deputy Associate Administrator for Planning in Washington, stepping down two years later. With his death in 1977, space exploration lost one of its most prominent pioneers.

HUNTSVILLE – ROCKET CITY

In April 1952, von Braun moved to a new test facility at Huntsville in Alabama. Thanks to the success first of the missile programme and then of the lunar launch vehicles, this little town of 16,000 inhabitants burgeoned into a metropolis with a population of more than 100,000, winning itself the nickname of Rocket City. In 1960 the missile centre, which had previously been operated by the US army, came under the control of NASA. Von Braun was director of the new Marshall Space Flight Center for the next decade.

Lift-off
The maiden launch of the Redstone missile from Cape Canaveral in Florida on 20 August, 1953, was von Braun's first major US success.

APOLLO 11 MISSION – 1969
First steps on the Moon

Cold War rivalries set off a race between the USA and the USSR to put the first men on the Moon. Although they initially lagged behind, the Americans ended up the clear winners. On 21 July, 1969, an estimated 530 million television viewers watched in astonishment as two men stepped out onto the lunar surface.

Launch vehicle
The Saturn V rocket (left), constructed at the John F Kennedy Space Center on Merritt Island, Florida, was the launch vehicle used for the Apollo programme.

Space pioneers
Seen here under assembly, the Mercury capsule was used in the early days of the US space programme for orbital flights designed to study the effects of weightlessness. It could only house a single astronaut.

'One small step for [a] man, one giant leap for mankind.' Neil Armstrong, the commander of the Apollo 11 mission, spoke the words as his foot touched down on the Moon's surface, leaving a clear imprint in the fine dust of the Sea of Tranquillity. His crackly tones took barely two seconds to travel the 384,000km (239,000 miles) to Earth. Buzz (Edwin) Aldrin followed soon after. The third member of the Apollo 11 crew, Michael Collins, was orbiting patiently in the command module 100km (60 miles) above their heads.

Those steps marked a decisive victory for the USA in the race to the Moon, fulfilling the promise made by President John F Kennedy eight years earlier when he had announced in

1961 '... the goal, before this decade is out, of landing a man on the Moon and returning him safely to the Earth'. For NASA and the 400,000 or so individuals who had taken part in the adventure, it was a moment to savour after a period of concentrated technological innovation undertaken at breakneck speed.

A massive launch vehicle

The *beep-beeps* of the Sputnik satellites had set off the space race in 1957. The affront to US national pride led first to the creation of the Mercury programme, whose goal was to put an American in orbit around the Earth. But once again, the USSR got there first. In April 1961 Yuri Gagarin became the first man in space when he made his historic Earth-orbit flight. Three weeks later Alan Shepard became the first American in space, but for just five minutes on a suborbital trajectory.

It was February 1962 before the first American finally went into orbit, and even then John Glenn spent only a quarter of an hour there. Meanwhile, the Soviet Semiorka rockets were superior to anything that the American space programme had to offer. The Mercury capsule that carried Glenn weighed a mere 1.5 tonnes, and it was accepted that a Moon launch would involve the equivalent of putting a 118-tonne payload in low Earth orbit.

The Saturn programme was created to meet the challenge, under the direction of Wernher von Braun. The prototype Saturn I rocket was powered by eight H-1 engines, fuelled with kerosene and liquid oxygen.

> ### THE SOVIET PROGRAMME
>
> While NASA concentrated on catching up technologically with the USSR, Soviet scientists also had their eyes on the Moon. But the authorities made the mistake of setting up two rival teams. One, under Valentin Glushko, worked on the Proton rocket, designed to enter lunar orbit, while the other, directed by Sergei Korolev, developed the heavy N1 craft conceived to land Russian cosmonauts on the Moon's surface. But the Proton was regarded as insufficiently reliable, while explosions on the N1 in 1969 and again in 1971 and 1972 effectively put paid to the Soviet plans. Even so, the unmanned Luna and Lunokhod programmes succeeded in bringing photographs and soil samples back to Earth.

On their way
A crowd assembled at Cocoa Beach (below), near Cape Kennedy (now Cape Canaveral) on the morning of 16 July, 1969, to watch the launch of Saturn V carrying the Apollo 11 astronauts. The launch went well, blasting the rocket into the sky (right).

APOLLO 11 MISSION – 1969

Apollo 11's crew
From left to right: Neil Armstrong, Michael Collins and Edwin 'Buzz' Aldrin. Only Armstrong and Aldrin touched down on the Moon's surface in the Eagle lander (bottom); Collins remained in the orbiting command module.

The first flight took place on 27 October, 1961, reaching an altitude of 137km (85 miles). The goal then became to build a more powerful rocket – one that was almost twice as long. It was achieved with Saturn V, first used on the Apollo 6 mission in 1968. Equipped with five F-1 engines, still burning the same fuels, this monster stood 111m (363ft) high; the first stage alone was as tall as a ten-storey building. To house it NASA built a gigantic hangar, the Vehicle Assembly Building or VAB, that was big enough to shelter five of the Saturn V rockets.

Designing the craft

North American Aviation had been responsible for the second stage of the Saturn launch vehicle, and they were also given the job of designing the spacecraft itself. It was a huge challenge: three men would have to spend eight days in the module in the most hostile environment imaginable. There was also a built-in dilemma, in that the larger and more commodious the capsule, the heavier it would be, making its return into the Earth's atmosphere more perilous.

The solution consisted of creating two separate units: the service module, containing essential resources that would burn up on re-entry into the atmosphere, and a command module in which the astronauts would live and touch down at journey's end.

The command and service modules were first flight-tested in 1966. It was in the first of these that NASA suffered the worst tragedy of the Apollo programme. On 27 January, 1967, the three-man crew were rehearsing just three weeks before the planned first manned flight when fire swept through the craft, killing 'Gus' Grissom, Ed White and Roger Chaffee. In the wake of the disaster the module had to be substantially redesigned, and 22 months passed before a crew finally went into orbit.

Planning the landing

From the start of the Apollo programme the most difficult question of all was deciding how to achieve touchdown on the Moon's surface. The concept of a dedicated lunar module, separate from the command and service craft, eventually won out over other suggestions, but the sheer weight of the proposed payload forced the designers to divide it into two stages: a 10-tonne descent stage, which would be abandoned on the Moon, and a 5-tonne ascent stage that would return to the orbiting mother craft with the astronauts in it. The Grumman Aircraft Corporation, which designed the module, decided to do without any pump or ignition system, relying instead on two tanks filled with nitrogen tetroxide and hydrazine, which combust spontaneously when mixed.

A series of successful launches followed the 1967 disaster, putting the programme back on track. In 1968 the Apollo 7 mission finally put a crew in space. Apollo 8 gave humans their first sight of the dark side of the Moon. Apollo 9 also launched and returned without problems, preparing the way for Apollo 10 in May 1969, which became a full-scale dress rehearsal for the planned Moon landing. Astronauts Tom Stafford, Gene Cernan and John Young did everything but touch down on the lunar surface. The next mission would be the real thing.

'The *Eagle* has landed'

On 16 July, 1969, the 130 million horsepower first-stage engine lifted the white Saturn V rocket off the ground. Three and a quarter hours later all three stages had been jettisoned, leaving the *Columbia* command and service modules, with the *Eagle* Moon lander attached, to travel on to the Moon. Three days later the craft was passing behind the Moon, making

THE SMELL OF MOON DUST

When they removed their helmets on their return to the command module, the two astronauts were struck by a smell like 'wet ashes in a fireplace' (Armstrong's words), or to quote Aldrin, 'spent gunpowder'. So the Moon has a distinctive odour – that of the soil covering it, which the men brought back with them on their boots. Like powdered graphite, the dust proved to be very abrasive and even allergy-inducing – a serious concern if a lunar base is ever established.

corrections to its orbit. On 20 July, Armstrong and Aldrin entered the lunar module, which separated from *Columbia* to make the descent to the Moon's surface.

The most delicate phase of the mission now got under way: the lunar descent. Controlled by an on-board computer, retro-rockets braked the craft and helped to stabilise it. At this nervous time a yellow alarm light blinked on in the cockpit; as it turned out, the computer was simply suffering from information overload.

A second cause for concern was that the trajectory of the descent was carrying the craft toward a crater. Taking over semi-manual

A NEW VOCABULARY

Astronaut, cosmonaut taikonaut, spationaut . . . all words meaning essentially the same thing, although with different nuances. The first three are used, respectively, for Anglo-American, Russian and Chinese space travellers, while the fourth has been provisionally adopted for those involved in manned flights of the European space programme.

Small steps

Neil Armstrong took this photograph of Buzz Aldrin (above). The lunar surface is covered in regolith, a layer of soil formed from meteorite impacts that are evident in the photograph of the dark side of the Moon (inset) taken by Michael Collins from the command module.

THE SCIENTIFIC LEGACY

The Apollo 11 astronauts took with them a solar wind collector, a seismometer and a laser reflectometer, designed to check the exact distance between the Earth and the Moon. Rocks were collected on all the missions involving lunar landings. Yet the Apollo programme was not conceived with science in mind – an omission whose effects only became apparent over the course of time. The crop of rocks – some 380kg (840lb) in all – was not even sufficient to settle once and for all the question of how the Moon was formed. In general the scientific legacy was sparse considering the cost of the programme, which amounted to between $20 and $25 billion, equivalent to almost $150 billion today. The technological advances were more apparent, including the development of new alloys and composite materials, as well as ultra-precise instrumentation and, above all, computer technology and integrated circuits, for which NASA was the world's main customer at the time.

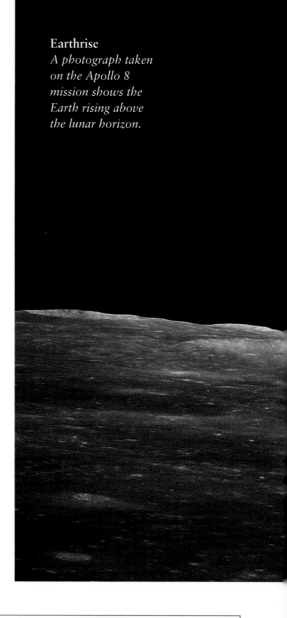

Earthrise
A photograph taken on the Apollo 8 mission shows the Earth rising above the lunar horizon.

Splashdown
The Columbia *command module containing the Apollo 11 astronauts successfully landed in the Pacific Ocean on 24 July, 1969 (left). The crew spent the next 21 days in quarantine to check that they had not brought any dangerous viruses back from space. President Richard Nixon visited them there.*

control, Armstrong brought the *Eagle* down 7km (4.3 miles) from the planned landing site. It was a close-run thing: less than 20 seconds' worth of fuel remained.

Once the craft was down, Armstrong spent two and a half hours on the lunar surface and Aldrin 90 minutes, collecting samples and taking photographs. Having repressurised the cabin and rested, the two men then activated the *Eagle*'s ascent stage to return to *Columbia*, bringing with them 23kg of lunar rocks. The *Eagle* was abandoned in lunar orbit.

On 24 July, having jettisoned the service module, the astronauts prepared to re-enter Earth's atmosphere on board the small command module. Braked by parachutes, the craft came down in the Pacific Ocean, where the aircraft carrier *Hornet* was waiting to meet it. After three weeks in quarantine in a specially converted trailer to ensure that no lunar micro-organisms could spread contagion on Earth, the three men came out to enjoy their moment of glory.

SAVED BY THE LUNAR MODULE

In April 1970, some 55 hours into the Apollo 13 mission to the Moon, a spark ignited an explosion in an oxygen tank in the service module. The crew of three – Jim Lovell, Fred Halse and Jack Swigert – took refuge in the cramped *Aquarius* lunar module, whose descent-stage rockets allowed them to correct the craft's trajectory to swing around the Moon and head back toward Earth. The module had been designed to house two astronauts for 48 hours, but had to shelter all three crew for four days. It proved a life-saver. Grumman, the firm that built it, sent the command module's manufacturer a bill for $312,421.24, jokingly labelled for 'towing fees'.

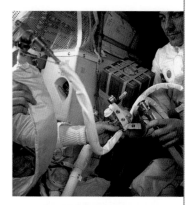

Improvised life-savers
Jim Lovell and Jack Swigert prepare makeshift filters designed to reduce the dangerous carbon dioxide build-up in the lunar module of Apollo 13.

APOLLO 11 MISSION – 1969

Lunar rover
The solar-powered all-terrain vehicle (inset, above right) was built for the Apollo 15 mission. It had a maximum speed of 8mph (13km/h).

An uncertain future

Officially, the Apollo 11 mission lasted for 8 days, 3 hours, 18 minutes and 35 seconds. Afterwards six other missions (Apollos 12 to 17) also headed for the Moon. Three more were planned, but the US Congress cancelled them on budgetary grounds. By that time US superiority in space was an established fact, and the hour of détente had arrived.

In 1972 the Apollo 17 crew spent three days on the Moon, undertaking three separate geological excursions in the course of which they collected 50kg (110lb) of rocks and dust and carried out dozens of experiments that would subsequently provide years of work and analysis for Earth-bound scientists. Eugene Cernan was the last to leave the lunar surface, on 17 December, after the team had between them put in more than 22 hours of extra-vehicular activity.

Will people return to the Moon one day? Former US President George W Bush indicated that they would when he launched the Constellation programme in 2004, proposing a new Moon shot in about the year 2020. But in the wake of the ensuing global financial crisis, President Barack Obama decided in February 2010 to abandon the project, preferring to prioritise manned flights to the asteroid belt, which could set the scene for an international venture to Mars.

Meanwhile new players are coming on stage. The Chinese, who have already had some success with manned orbital flights, have announced a Moon programme that envisages a landing around the year 2020. India has yet to reveal the scope of its ambitions.

A SCIENTIST ON THE MOON

All the Apollo astronauts were present or former members of the US armed forces with one exception: Harrison Schmitt. Born in 1935, Schmitt was the only scientist in the astronauts' ranks, having studied geology before joining NASA. On the Apollo 17 mission, Schmitt collected the rock labelled Troctolite 76535, which is now judged, for its age and state of conservation, to be the most interesting of all the samples brought back from the lunar surface.

INVENTIONS THAT CHANGED OUR LIVES

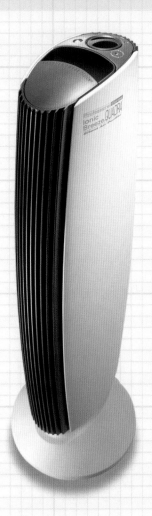

The ioniser 1969

Invented in 1969, the air ioniser is a device that uses a high-voltage electric charge to generate a flow of negative ions. These link up with positively charged particles in the atmosphere to purify the air. Dust, microbes, pollen and other pollutants are filtered out and disposed of by a particle collector.

There are also water ionisers designed to treat drinking water, which their supporters claim have antioxidant properties: some are used to purify the water in swimming pools. Ionisers are also employed in certain air-conditioning systems.

BUT DO THEY WORK?

The effectiveness of ionisers has often been questioned. In the USA, *Consumer Reports* magazine maintained, on the basis of tests it had carried out, that they were less efficient than conventional air filters. Yet a study published in *New Scientist* in the same year, 2003, reported spectacular success in reducing airborne infections in a Leeds hospital ward. The debate continues.

The POSM 1969

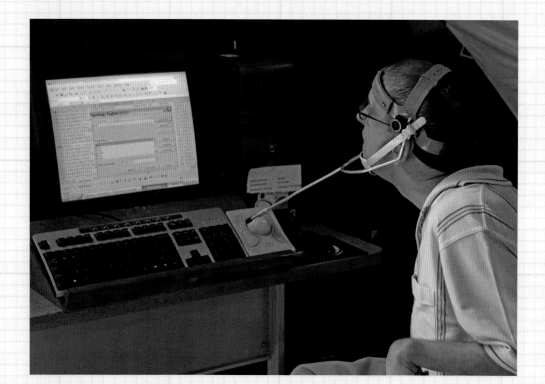

The Patient-Operated Selector Mechanism (POSM) – popularly referred to as the Possum – is a control device that enables severely handicapped people to perform specific functions and communicate.

Developed by Reg Maling and Derek Clarkson at Stoke Mandeville Hospital in Aylesbury, Buckinghamshire, POSM first saw the light of day in 1969. It enabled patients – including those who were paralysed – to activate commands giving them control over various electrical appliances. In the early days this was by means of a breathing tube. The development of a 'sip-and-puff' typewriter allowed them to communicate their thoughts; one individual even managed to compose a book of poetry.

By the mid 1980s computerisation had extended the range of possibilities; users equipped with remote controls could perform a wide variety of activities that included making telephone calls and playing chess. Portable units have been available since 1993, programmed to meet individual needs.

Disabled enabler
The POSM system allows severely handicapped people not just to write and receive messages but also to control bells, lights, radios, telephones and televisions.

Implantable defibrillator 1970

In 1966 Polish-born cardiologist Michel Mirowski suffered the loss of an old friend to a heart attack brought on by ventricular fibrillation – uncoordinated contractions of the cardiac muscles. An electric shock to the heart could have revived him, but the defibrillation procedure, first used on humans by Claude Beck in the USA in 1947, required the use of a cumbersome apparatus. Mirowski decided to develop an implantable defibrillator that would do the job within the body itself.

Featherweight

External defibrillators deliver an electric shock of some 300 joules through the rib cage. Mirowski believed that the discharge strength could be reduced by a factor of 10 if the stimulation was delivered directly to the heart. In 1970, with the American cardiologist Morton Mower, he implanted in a dog a prototype device, consisting of a programmable pulse generator linked by electrodes to the heart to monitor the cardiac rhythms and deliver shocks when required. The first human implantation followed in 1980, when a device weighing about 300g was lodged in a woman's abdomen. Later versions were even lighter, about 80g, while their working life was extended from 18 months to six or seven years.

In the years that followed, new functions were added. Besides defibrillating, the devices also came to serve as cardioverters (providing low-intensity shocks to counter tachycardia), Holter monitors (constantly analysing heart activity over a period of time) and pacemakers. Today ICDs (implantable cardioverter-defibrillators) are used for patients presenting various forms of life-threatening cardiac arrhythmia who cannot be treated with drugs. A 2008 study covering 15 European nations showed an average usage of 140 per million inhabitants, a figure that had risen rapidly over the preceding five years.

X-ray view
Located, like a pacemaker, under the skin on a level with the rib cage (above), this ICD is connected to a computer, so the information it provides can be monitored.

LIFE-SAVERS

External defibrillators have also benefited from miniaturisation and technological advances. Automated devices that diagnose ventricular fibrillation by monitoring heart rhythms became available from 1994 on, programmed to deliver shocks of between 150 and 350 joules. Recent models can be used by anyone and are sometimes located in public places, like this one (left) in a railway station in Lyon, France. Estimates suggest that their widespread use could increase the survival rate for people suffering heart attacks from 3 or 4 per cent to more like 30 per cent.

CIVIL AVIATION
Air traffic takes off

In the 1960s and 70s, air transport moved into high gear, emerging as a mass-market industry. Jet propulsion became the norm in civil aeroplanes, and demand from the public soared. The economic and cultural effects were far-reaching.

Counter service
A passenger checks in at Orly Airport outside Paris in the 1970s.

In October 1958 two revolutionary jet airliners crossed the Atlantic on pioneering commercial flights. The first, on 4 October, was a Comet V operated by the British Overseas Airways Corporation (BOAC); the second was a PanAm Boeing 707. Carrying twice as many passengers as the Comet, at a cruising speed of around 560mph (900km/h), the Boeing had both a technological and commercial edge over the British plane. That edge would become a yawning gap following a couple of disastrous early accidents involving Comets, whose reputation never recovered. But for now, the flights confirmed that a new era was opening in civil aviation – one that saw the major airlines turn their backs on piston engines and propellors. Air travel at the time was the preserve of the privileged, who now became known as the 'jet set'.

Boeing leads the pack

Despite resistance from the American-built Douglas DC-8, which entered service in 1959, Boeing quickly came to dominate the market. The 707 soon had fresh rivals. From 1959 there was the Convair 880, followed by the Convair 990 from 1961. Over the course of the 1960s came the Lockheed Tristar and Douglas DC-10, as well as the British Vickers VC10 and BAC One-Eleven, but none really threatened Boeing's predominance. The French developed the Caravelle (1959) as a short and medium-range jetliner not equipped for transatlantic flights.

LEADING THE WAY BEHIND THE IRON CURTAIN

On 22 March, 1956, the Soviet leader Nikita Krushchev arrived in London on a state visit aboard a Russian jetliner. Westerners looked agog at the Tupolev 104, which resembled a Comet but had the engines set tight up against the fuselage at the base of the wings (the Russian plane only had two jet engines compared to the Comet's four). Developed from the Tu-16 Badger strategic bomber, the Tu-104 was noisy and narrow-bodied, but could carry between 50 and 70 passengers (later, up to 120) at over 500mph (800km/h). For two years after it came into service between Moscow and Irkutsk, the Tu-104 was the world's only commercial jetliner. It met a need, for the vast extent of the Soviet Union made rapid air transport desirable. As a result the Soviet state-owned Aeroflot was for many years the largest airline in the world in terms of passenger numbers carried. Comfort was never its priority, but the Tupolevs and Ilyushins that it flew were effective workhorses. The collapse of the Soviet Union in 1991 caused a decline in its civil aviation industry that Russia is currently trying to reverse.

Trailblazer
A Tupolev 104 on the tarmac at Moscow's Sheremetyevo Airport in 1959.

Euroliner
The Airbus A300 was produced by a European consortium, entering service between London and Paris in 1974.

On 18 December, 1970, the French company Aérospatiale and Germany's Deutsche Airbus signed an agreement that formed a new cross-border consortium called Airbus. The first Airbus A300 made its maiden flight in 1972. Subsequently the company would come to rival Boeing in the number of aircraft sold, even overtaking the American company in the opening decade of the 21st century. The duel between these two giants gradually eliminated all the other competing long-range constructors, who one after another disappeared.

Unceasing progress

To a casual observer, today's airliners look surprisingly little different from those of the 1960s; they have the same general outline, including the positioning of the engines. But what has changed is vitally important. To start with, the engines themselves have become much less noisy and consume less fuel. The planes are also safer: in 1987 the Airbus A320 became the first commercial airliner to have a fully electronic fly-by-wire system, which meant that the pilot no longer needed to control the plane manually; instead, his instructions were transmitted as electronic impulses via computers to actuators manipulating the controls. On Airbus jets, a small lever resembling a joystick took the place of the traditional control column. Piloting thereby became more exact and fuel-efficient, while the computers serving as intermediaries

Nosing along
The DC-8, shown here head-on (below), was a four-engined medium-to-long-range jetliner made by the Douglas Aircraft Company from 1959 to 1972. The prototype first flew in May 1958.

Economy class
Tourist-class passengers, like these on a Boeing 747 in the 1970s, were given a more basic service than first-class travellers, who on some flights already had access to a small bar.

AN INTERNATIONAL REGULATOR

In November 1944 a conference in Chicago brought together air-transport authorities from around the world to put their signatures to a joint convention: the International Civil Aviation Organisation (ICAO) was born. A specialised agency of the United Nations with headquarters in Montreal, it has responsibility for the security and reliability of air travel, setting standards for aeroplanes, airports, air traffic, staff training and other related matters.

CIVIL AVIATION

GOING SUPERSONIC

With a roar the great, white, delta-winged plane took to the sky, its silhouette sharp as an arrowhead. The date was 2 March, 1969, the place Toulouse in France, and Concorde 001 was making its maiden flight. The Concorde story had begun back in November 1962, when the British Aircraft Corporation (BAC) and France's Aérospatiale, each supported by their governments, agreed to jointly develop an airliner capable of carrying passengers across the Atlantic at twice the speed of sound. The plane they delivered was a technological triumph which made its first transatlantic flight in 1973 – the year of the first oil crisis. Concerns about the plane's high fuel consumption, as well as environmental worries about the sonic boom it created, led international orders for Concorde to drop away. Yet commercial flights with British Airways and Air France began three years later. Concentrating on profitable routes, from London or Paris to New York or Washington for example, it continued in service until 2003. But high maintenance costs and the loss of a plane in France in 2000 – the only fatal accident in Concorde's history – finally sounded the death knell of the only supersonic plane to deliver commercial passenger services.

between the pilot and the controls introduced new flight-management possibilities. The vast array of dials, buttons and other devices that had cluttered earlier cockpits gave way to a handful of computer screens. At ground level, first computerisation and then the Internet revolutionised the business of booking tickets.

Mass-market travel

In 1947 airlines carried 21 million passengers around the world. By 2006 that figure had increased by a factor of 100 – that year 2.1 billion people embarked on scheduled flights. The two figures do much to explain why it became necessary to rethink aircraft design in the intervening years. In the early 1970s two competing trends came to the fore: greater speed, so that more return journeys could be made each day, and greater size, so that planes could carry more passengers. In 1969 two iconic planes made their maiden flights: the Anglo-French Concorde, which flew at twice the speed of sound, and the Boeing 747, the first jumbo jet carrying from 350 up to 600 passengers. Concorde symbolised speed and luxury, the 747 mass travel – and it was the 747 that won out. Despite initial hostility to the 'jumbo' concept, other large-capacity aircraft with seating for 300 to 400 passengers were introduced from the early 1970s on, including the Lockheed Tristar, the McDonnell-Douglas DC-10 and the Airbus A300.

In-post training
Air France pilots practising on a flight simulator at a training facility at Massy in the Parisian suburbs (below). The control panel predates computers.

Cutting costs

Meanwhile, in-flight service was also changing. In the early days only the wealthy could afford to fly and there was only one class of seating on board airliners. Then tourist class appeared on scheduled flights. More radically, charter flights began in the mid 1970s. A tourist agency or a club would take a block booking on a scheduled flight or rent an entire aircraft and then resell the tickets, usually as part of a package that also included hotel reservations. Initially charters had a bad reputation, as passengers rarely knew which airline would carry them, but that soon changed.

Established airlines responded by offering special tariffs, preferring to cut profit margins rather than dispatch empty planes. Low-cost airlines appeared in the late 1990s in the shape of Ireland's Ryanair and Britain's EasyJet, offering scheduled flights that cut costs in all departments except aircraft maintenance and staff training, both of which were governed by international regulations.

At the same time air freight transport has grown in similar fashion. When the holds of airliners no longer provided sufficient room to carry it all, manufacturers began producing special versions of their planes designed purely as cargo-carriers. The Boeing 747-8 could shift 140-tonne loads over 5,000 miles (8,000km), the Airbus A330-200F could carry 69 tonnes for 3,700 miles (6,000km). As a result, goods were being transported around the planet faster than ever before.

Ups and downs

The story of civil aviation has directly reflected the ups and downs of the wider world. The steady improvement in quality of life in developed countries since the 1950s sent demand soaring. The oil shocks of the 1970s put paid to the first generation of jet airliners, which were too greedy of fuel, forcing airlines to rethink their pricing structures in order to keep their planes filled. Another factor has been terrorist acts such as

Transport hub
Frankfurt Airport (above) is the largest in continental Europe. It is also one of the busiest; in 2006 it handled more than 12 million passengers.

Operations hub
More than 260 air traffic controllers operate the terminals at Charles de Gaulle Airport near Paris (above), working from three control towers to maintain traffic flow and guide the planes in to land.

the bombing of a PanAm 747 over Lockerbie in Scotland in December 1988, which led to new security measures. More recently, the terrifying attacks on planes in the USA in September 2001 had serious implications for the entire industry. Stringent new security checks on passengers have lengthened the check-in process in airports. Air travel has also been blamed for the spread of viruses, such as the SARS epidemic of 2003 or the bird flu and swine flu outbreaks of 2004 and 2009. The economic turmoil at the end of the decade and the growing threat of global warming continue to provide fresh challenges for plane operators and constructors.

Tight fit
A Boeing 747 comes in to land at Kai Tak Airport in Hong Kong, a demanding airport for pilots who had to steer a course between mountains and tower blocks. It was replaced by a new facility in 1998.

Natural hazard
The eruption of Eyjafjallajökull (below), a volcano in Iceland, created unprecedented disruption to air traffic across northern Europe in April 2010. The losses for airline companies were estimated at almost $2 billion over a two-week period.

TWIN THREATS

With the exponential growth of air traffic, civil aviation now faces two major challenges. On the one hand, the world's existing airports are reaching the limits of their capacity and the construction of new ones often runs up against local opposition – and this at a time when the sheer volume of traffic is posing increasing problems for flight controllers, whose job is to maintain and if possible improve standards of security. On the other, in spite of the progress made in designing more efficient jets, airliners still use huge amounts of fuel, a limited resource that is constantly becoming more expensive.

Animal expert
A trained Alsatian sniffer dog checks baggage for narcotics or explosives (below).

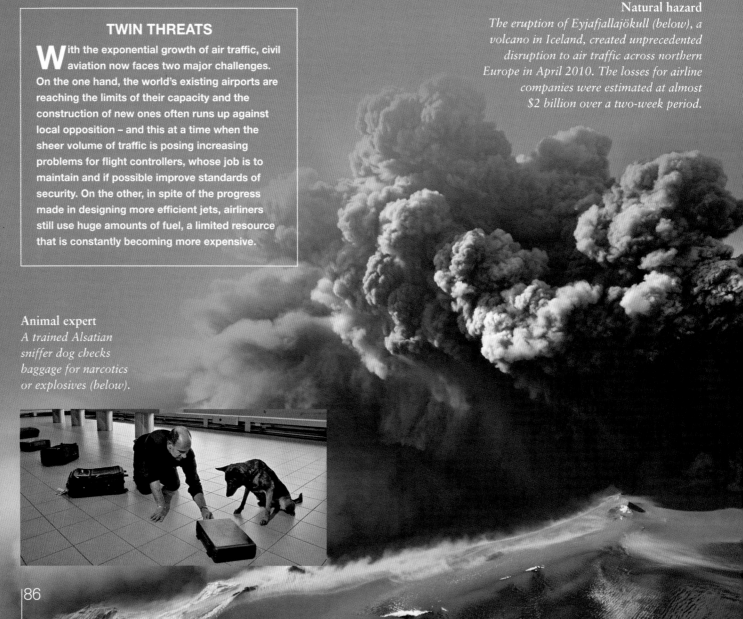

INVENTIONS THAT CHANGED OUR LIVES

The water jet cutter 1971

Several new composite materials had been developed over the course of the 20th century, but what tools were best adapted to cut them so as not to damage them in any way? The question was an important one, because the heat generated by metal cutting edges had a tendency to deform some materials, while others simply crumbled away. In the late 1950s a Canadian forestry engineer by the name of Norman Franz dreamed up a simple solution to the problem. In place of metal blades and mechanical saws, he proposed using a high-pressure water jet, with a force of between 100 to 700 bar, all concentrated onto a tiny surface area.

Franz built prototypes and took out his first patents in 1968. His system worked, but only on relatively soft materials like wood – when the pressure was increased to cut harder surfaces, the water instantly vaporised, rendering the jet useless. In 1968 Franz sought to solve this problem by adding carbon-reinforced polymers to the water in order to prevent vaporisation and to concentrate the jet even more.

A tool for industry

The first industrial application of the new technology came in 1971, when the US firm Alton Box Board developed a cutting tool with a pressure of up to 4,000 bar that was accurate to within just half a millimetre.

In 1985 a US entrepreneur named Michael Pao had the idea of adding solid particles to the water jet, making it abrasive enough to cut really tough materials – the only exceptions being diamonds, tempered glass and certain ceramics. Today water jet cutters are used not just in the manufacture of metal sheets and composites, but also in the food industry for slicing fruits without contaminating them and in the construction industry to strip concrete without causing cracks.

High-speed erosion
Mounted on a bracket, the cutter operates like a bandsaw. The material to be cut – here, a block of aluminium – is arranged beneath the jet by hand.

Force of water
Water from a high-pressure inlet is fed at a pressure of 4,000 bar through a sapphire or diamond nozzle that concentrates the flow into a fine jet just 0.1mm to 0.3mm across, travelling at just under 1,000m per second – almost three times the speed of sound.

A HIGH-PRESSURE MEDIA COUP

Michael Pao founded Flow International in 1970 to produce high-pressure water jet cutters. He sought to sell the machines to Kimberly-Clark, the principal US-owned manufacturer of articles of personal hygiene, but was initially rebuffed. Pao's next step was to place an advertisement in the press, ironically referring to the firm's openness to innovation. Stung by the jibe, the multinational tested Pao's equipment and a few months later installed it in all its factories.

THE MICROPROCESSOR – 1971
More power to the chip

The invention of the microprocessor marked a decisive step forward for the electronics industry. Highly sophisticated integrated circuits paved the way for the construction of the first personal computers by placing all of their logic gates onto a single chip. The revolution began with the development of the Intel 4004.

In 1969 Intel, a young company based in California's Silicon Valley specialising in the production of electronic components, accepted a commission from a Japanese calculator manufacturer, Busicom, which wanted to put a dozen integrated circuits on a silicon substrate. Intel gave the job to Marcian 'Ted' Hoff, a specialist in the miniaturisation of logical systems. In 1971 Hoff succeeded in fitting all the processing functions onto a single chip, shifting the ancillary circuits to three other support chips. Dubbed MCS-4, the family of chips based on this technique included the 4004, the first microprocessor.

Toward personal computers
The Intel 4004 compressed a formidably dense array of components, jamming 2,250 transistors onto a single chip with a surface area of just 3mm². It was programmable and functioned as a universal integrated circuit that could take the place of more specialised electronic circuits. Designed to handle digital data, microprocessors are extremely complex circuits that operate to the rhythm of an internal clock, executing instructions at regular intervals; the 4004's instruction period was 10.8 microseconds.

To operate, microprocessors require the support of a variety of other units, including microsequencers, counters, processor registers, floating-point units, workflow systems for the inflow and outflow of data and an ultrarapid internal memory. The way in which these various elements are combined has, over the course of time, come to define the different types of processor. Miniaturisation has also been a feature of their development, increasing their capabilities even as they have physically shrunk in size.

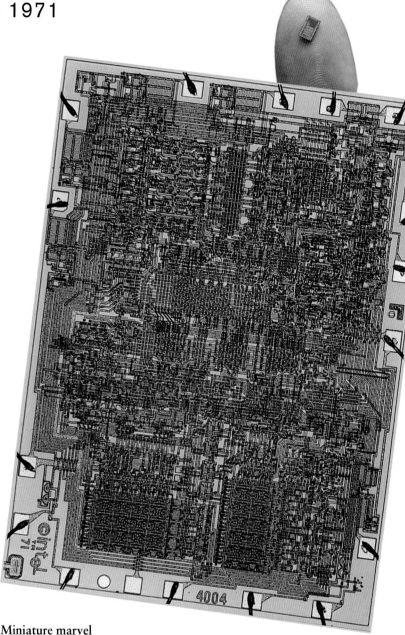

Miniature marvel
The working side of an Intel 4004 microprocessor (above), magnified more than 30 times. The actual chip, shown in relation to a finger tip (top,) measures 3.81 x 2.99mm.

DEFINITION
A microprocessor is an integrated circuit contained on a single computer chip that can carry out most of the functions of the central processing unit of a small computer.

A GENERIC TERM
Integrated circuits carry information in the form of electronic impulses, and microprocessors carry out operations by putting the integrated circuits to work. The two terms have become inseparable from one another and from the microchips that carry them, but there are integrated circuits that are not microprocessors.

THE MICROPROCESSOR – 1971

MOORE'S LAW AND ITS LIMITS

Californian-born Gordon Moore was a co-founder of Intel. His law claims that the number of transistors that can be inexpensively placed on a microprocessor will double every two years, increasing performance while costs remain static. The formula has held true since first stated in the 1970s, but researchers at Intel now think that it will reach its limits in 2018, when miniaturisation reaches a physical boundary: that of the size of atoms. While the smallest wires employed on the 4004 were about 10 microns in diameter (ten times less than a human hair), current microprocessors have reduced that figure to just 0.65 microns.

Making microprocessors
Microchip manufacturers have developed increasingly efficient techniques (above) that have steadily reduced the surface area of the processors' integrated circuits and increased their computing speed, while taking care to avoid the risk of overheating.

THE EVOLUTION OF MICROPROCESSORS

Date	Name of Microprocessor	Number of Transistors
1971	4004	2,250
1972	8008	3,500
1974	8080	6,000
1978	8086	29,000
1982	80286	275,000
1989	80486	1.2 million
1993	Pentium	3.1 million
1997	Pentium II	27 million
2000	Pentium	442 million
2006	Core 2 Duo	291 million
2007	Core 2 Quad	582 million
2008	Core i7	730 million
2010	Core i7 980X Extreme Edition	1.3 billion

Intel first produced the 8086 microprocessor in 1978; it was used in the IBM PC, the iconic personal computer introduced in 1981, paving the way for something approaching a monopoly hold over the home computer market. Motorola, which provided microprocessors for the early Apple Macintoshes, was its main rival.

Chips with everything

From that time on the power of computers came to seem limitless. The market was changing, as games and video applications were added to office uses, making huge demands on processing capability. Yet at the same time that microprocessors were gaining in complexity, each new generation was cheaper to produce, partly thanks to Moore's law (see above) but also due to reductions in manufacturing costs as volumes grew. Personal computers also grew steadily less expensive.

In the 1990s, microprocessors came to play an increasingly important role in the digital economy. Computers continued to shrink in size and the number of devices with processors inside them mushroomed. Microchips became an essential part of life, finding their way, ever more discreetly, into our cars, houses, telephones, toys, sports equipment – even the human body itself.

Chip package
An Intel 4004 microprocessor rests in its 16-pin carrier (left). It consists of a block of silicon containing tiny electronic circuits comprising 2250 separate transistors. Its edges harbour over 160 miniscule electric contacts. Instructions and data pass down the pins in the form of electronic signals.

SPACE TRAVELLERS
The stuff of heroes

In the half century since the first manned space flight in 1961, fewer than 500 men and women have gone into space. The heroic pioneers of the early days have given way to scientists and technicians – brave men and women nonetheless – living for extended periods in Earth orbit studying some of the problems posed by the challenge of space travel.

At 9.07am Moscow time on 12 April, 1961, the Vostok 3KA-3 spacecraft, later to be renamed Vostok 1, was launched in conditions of extreme secrecy from the Baikonur Cosmodrome in Kazakhstan. In the capsule at its tip was Yuri Gagarin, a former apprentice foundryman, now a 27-year-old fighter pilot with the Soviet Air Force, one of an elite group of pilots chosen to train for the space programme. He was about to write his name into the history books.

The flight was automated throughout, for no-one knew at the time how a human pilot would react to the condition of weightlessness, although there was an emergency key in case Gagarin needed to take control. Ten minutes after launch, the last stage of the rocket burned

First man in space
Yuri Gagarin's extraordinary exploit made headlines all around the world (above). His landing capsule came down in a Russian field (left); Gagarin himself had ejected and drifted down by separate parachute.

out and the craft went into orbit around Earth, some 190 miles (300km) above the planet's surface. Gagarin completed a circuit of Earth and braking started at 10.25am, above the west coast of Africa. The re-entry module housing Gagarin failed to separate completely from the equipment module, remaining attached by a bundle of wires. This caused the craft to gyrate dangerously as it re-entered Earth's atmosphere above Egypt, but fortunately the cables burned up, freeing the capsule and allowing it to resume its expected trajectory. At 10.55am, after a flight of 1 hour and 48 minutes, Gagarin ejected at a height of 7,000m (23,000ft) to

SPACE TRAVELLERS

WOMEN PIONEERS

After the first space flight made by a woman, the Russian cosmonaut Valentina Tereshkova (right) in 1963, another 19 years would pass before a second woman travelled in space. Moscow-born cosmonaut Svetlana Savitskaya travelled to the Salyut 7 space station in 1982, and became the first woman to undertake a space walk. The first American woman in space was Sally Ride, a Stanford-trained physicist who made two flights in the space shuttle *Challenger*, in 1983 and 1984.

Spacewalker *Edward White (above) was the first US astronaut to venture outside a spacecraft. He spent a total of 15 minutes in the void during the Gemini 4 mission in June 1965, attached to the capsule by a tether. The first ever space walk had been taken earlier that same year, on 18 March, by Alexei Leonov on Voskhod 2.*

descend by parachute. He landed in a field in the Saratov district of southern Russia, to the astonishment of the farm labourers working there at the time.

This first manned flight had tested to the full the qualities required by the early space pioneers, whether Soviet or American. They need to be fearless adventurers, physically and mentally strong, but also technically skilled and able to improvise so they could, if necessary, come up with makeshift solutions to unforeseen problems. In the quest to find such people, the space agencies of both superpowers trawled the ranks of fighter pilots in their respective air forces seeking out exceptional individuals.

The ideal astronaut was someone of above-average physical condition, psychologically tough and endowed with a cool head and excellent reflexes. The only early recruit not from a military background was Valentina Tereshkova, who in 1963 became the first woman in space; she was chosen for her working-class background – ideologically important for the Soviets – and her experience as an amateur parachutist (see box above).

Survival lessons

In the early days of the space programme, the chief priority for astronauts in training was learning to live within the physical constraints of spaceflight. Scientific research skills came a distant second. Would-be space travellers had first and foremost to be able to survive in a hostile environment. Three weeks after Gagarin's flight, his US counterpart Alan Shepard went into suborbital space aboard a Mercury spacecraft, showing in his turn that the human body could resist weightlessness and the g-forces associated with rapid acceleration. As missions became longer, astronauts gradually became accustomed to the demands of life in orbit. More than half would later describe the chief problem as a space sickness, analogous to sea sickness, caused by weightlessness disturbing the body's natural sense of balance linked to the inner ear. The condition typically lasts for two or three days, causing nausea and in some cases also disorientation and visual illusions.

A second adaptation problem became apparent when the cosmonaut Alexei Leonov spent 10 minutes outside the Voshkod 2 craft on the world's first space walk. To survive, he had been equipped with a spacesuit modelled on suits worn by jet pilots exposed to extreme variations in temperature and air pressure. The suit was designed to provide protection both from the intense heat of the Sun's rays, which reaches 120°C (248°F) in the vacuum of space, and from temperatures that fall as low as -160°C (-320°F) in the shade. Pressurisation permitted the wearer to breathe. The problem with Leonov's suit was that the material proved so rigid when inflated that initially he could not get back through the airlock to rejoin his fellow-cosmonaut, Pavel Belyayev, inside the craft. After partly depressurising the suit he managed to squeeze through. The end

SPACE TRAVELLERS

SPACE SCOOTER

In 1984 Bruce McCandless (right) became the first person to test the Manned Maneuvering Unit (MMU), which allowed astronauts to move freely in space without being tethered to a spacecraft. While doing the job of a scooter, it looked more like a floating armchair, with a propulsion system fitted to the back and controls at the tips of the armrests. On Earth the unit weighed 148kg (over 23 stone). A lighter kit called the SAFER ('Simplified Aid for EVA Rescue'), designed as a backpack, was first tested in 1994.

SPACESUITS

Astronauts wear spacesuits as a safeguard against the risk of depressurisation, but mostly for extravehicular spacewalks. There, in the vacuum of space, they have to cope with extreme temperatures and also the risk of exposure to micrometeorite hits and the effects of cosmic rays. For protection, the suits are made up of several layers of different materials. Much time was spent on the design of the helmet, whose carbon-fibre visors, offering a combination of security and clear vision, are technological marvels. The suits weigh in at as much as 100kg (over 15 stone) on Earth, but are weightless in the zero-gravity conditions of space.

Consumer test
Coca-Cola prepared this experimental package for the joint US-Russian STS-63 space mission in February 1995. The aim was to test the effects of weightlessness on the sense of taste.

of the trip was possibly even more dramatic: after re-entry, they came down in a snowbound forest in the Ural Mountains, where the two men had to wait for a whole night, surrounded by wolves, before they were tracked down.

Not all astronauts were so lucky. In space exploration the least malfunction can prove fatal, as a catalogue of disasters has tragically revealed. The Apollo 1 fire killed three US astronauts on the launch pad in 1967 and two space shuttles have been destroyed by explosions: *Challenger* shortly after take-off in 1986 and *Columbia* on re-entry in 2003. In each case the entire crew of seven was killed.

Objective Mars

After the Moon landings and the cancellation of the Apollo programme, the adventure of space travel took a different turn. The objective now became to spend increasingly long periods outside the Earth's atmosphere in preparation for the launching of even more distant missions, for instance to explore the planet Mars. A new kind of astronaut came to the fore as jet pilots gave way to scientists, trained to study the reactions of their own minds and bodies under the rigours of space travel. Space stations were the laboratories for these experiments in endurance.

A new discipline of space medicine evolved to deal with the physical and psychological problems associated with life outside the Earth's atmosphere. Experience showed that weightlessness and the prolonged inactivity involved in life in the narrow confines of a spacecraft led to a loss of muscle mass, so

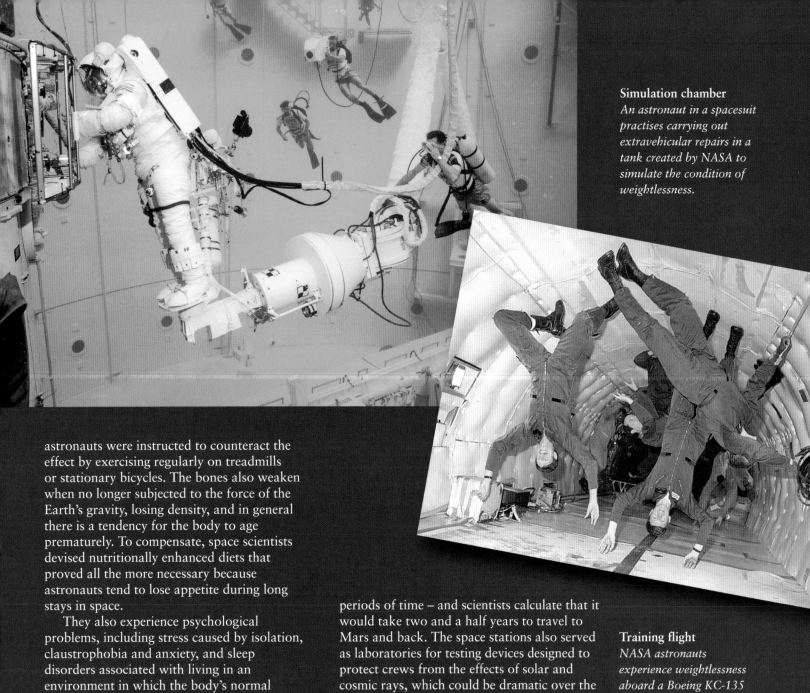

Simulation chamber
An astronaut in a spacesuit practises carrying out extravehicular repairs in a tank created by NASA to simulate the condition of weightlessness.

Training flight
NASA astronauts experience weightlessness aboard a Boeing KC-135 Stratotanker, adapted to simulate zero-gravity conditions (above).

Coping with g-forces
This centrifuge in the German city of Cologne is designed to test astronauts' resistance to acceleration pressures.

astronauts were instructed to counteract the effect by exercising regularly on treadmills or stationary bicycles. The bones also weaken when no longer subjected to the force of the Earth's gravity, losing density, and in general there is a tendency for the body to age prematurely. To compensate, space scientists devised nutritionally enhanced diets that proved all the more necessary because astronauts tend to lose appetite during long stays in space.

They also experience psychological problems, including stress caused by isolation, claustrophobia and anxiety, and sleep disorders associated with living in an environment in which the body's normal biorhythms are disturbed. These can lead to tension between crew members forced to endure one another's company over long periods of time – and scientists calculate that it would take two and a half years to travel to Mars and back. The space stations also served as laboratories for testing devices designed to protect crews from the effects of solar and cosmic rays, which could be dramatic over the course of long-distance voyages.

Such experiments, combined with others carried out on Earth in conditions simulating space, are all helping to prepare the way for an eventual mission to Mars. The Red Planet continues to haunt the imagination of the space-faring nations, new as well as old, for rising powers are also joining this new space race. In 2003 Yang Liwei became the first taikonaut (literally, 'space sailor'), as China's astronauts are called, entering Earth orbit in the Shenzhou 5 spacecraft.

SPACE STATIONS – 1971
Living and working in space

Once the race to the Moon was over, the rivalry between the USA and USSR found a new focus in near-Earth orbit. From the first Salyut module to the full complexity of the ISS, space stations have evolved hugely, while spiralling costs and the collapse of the Soviet Union transformed initial competition into international cooperation.

On 19 April, 1971, the Soviet Union put the first space station, Salyut 1, in orbit. Designed for long-stay manned flights, the craft was unlike anything that had gone into space before: it was 15m (50ft) long and comprised various compartments designed for different purposes – relaxation, maintenance and conducting scientific experiments. Four solar panels provided its energy requirements. Photovoltaic cells were a new technology at the time, but their efficiency was never called into question and they became a regular feature of subsequent space station designs.

The first crew reached the station a few days later, but a defect in the docking mechanism on their Soyuz spacecraft meant that they could not get inside. A second attempt in June was successful, and the three cosmonauts went on to spend 22 days on the station, sharing its 90m³ of living space. But the story ended tragically; on their return to Earth a pressure-equalisation valve in the re-entry capsule malfunctioned and all three men were killed.

Over the next 11 years six more Salyut research stations went into orbit, and the length of time that cosmonauts spent in them steadily increased, reaching 200 days or more. Their main task was to study the effects of weightlessness on the human body.

Meanwhile the USA launched Skylab in 1973. The programme was intended to study the Solar System along with the feasibility of long-range manned flights and was financed with the last of the Apollo mission money. It lasted barely a year. The empty craft eventually fell back to Earth in 1979.

The next major development came in Ronald Reagan's 1984 State of the Union address, when the US President announced plans for a big new space station to be called Freedom. It was assigned an initial budget of $8 billion and its remit included acting as a service station for satellites and operating as a zero-gravity laboratory. Other space agencies, notably those of Europe and Japan, were quick to offer support. But the project soon experienced delays as plans were changed and costs spiralled out of control, causing the launch date to be put back to 1992 at the earliest.

Space rendezvous
An artist's impression of a Soyuz spacecraft about to dock with the Salyut 1 space station. A hatch on the vessel opened onto the space station, permitting the cosmonauts to pass into the station without having to put on spacesuits.

ENDURANCE RECORDS

Russia's Valeri Poliakov holds the record for the longest continuous stay in space. In 1994 he spent 438 days – more than 14 months – on the Mir space station. His compatriot Sergei Avdeiev racked up a total of 748 days, but they were spread over three separate visits.

A new era begins

In this new space race the Russians once more had the edge on the Americans. In 1986 they put the first components of the big Mir space station into orbit. Over the next decade four more modules were added to the central core until, in its final form, the station stretched for 45m, weighed 140 tonnes and provided 400m³ of living space – enough to house up to ten cosmonauts at any one time. In the 15 years of its existence – it was originally planned to last five – Mir welcomed aboard 105 men and women of 11 different nationalities. In all, 110 different craft docked with it, some carrying astronauts but others bringing essential supplies including water, oxygen, provisions and equipment.

The station was first and foremost a big laboratory, in which a total of 31,000 experiments were carried out. The cosmonauts themselves were the main subjects, serving in effect as space guinea-pigs since the main focus of attention continued to be the study of life in conditions of weightlessness. In addition, work was done in the fields of animal and vegetable biology, astronomy, physics and materials science, all of it carried out 400km (250 miles) above the Earth's surface.

Yet in the long run the sum of knowledge gained from the experiments was a touch disappointing, adding up to little more than the lesson that humans could live in orbit for months at a time and return to Earth more or less in shape. After a time people began to

Test vehicle
Skylab (above), the first American space station, was launched on 14 May, 1973. Three manned flights travelled to it over a nine-month period. The last was the Skylab-4 mission, dispatched in November 1973, on which Edward Gibson made an extravehicular sortie (above right).

Construction kit
The Mir space station (below) was assembled from a number of modules dispatched separately into space. When complete, it was as long as six buses.

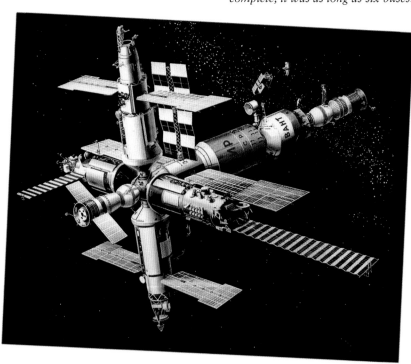

SPACE STATIONS – 1971

THE LAWS OF SPACE

In a crowded living space like the ISS, the possibility of conflict can never be ruled out, so a code of conduct has been adopted by all participating nations in an attempt to settle any problems that might occur. In principle each one has jurisdiction over, and responsibility for, the crew members and visitors it sends to the station, although some exceptional circumstances are allowed.

Historic handshake
Robert Gibson, commander of the Atlantis *space shuttle, greets Vladimir Dezhurov, his Russian counterpart, after docking with the Mir space station on 29 June, 1995.*

Joint mission
ISS crew members took part in a press conference in 2008, with images and sound relayed from the space station to the Korolev space centre near Moscow (above). The team aboard the ISS at the time was made up of Americans, Russians and a Korean.

15 YEARS OF GOOD AND FAITHFUL SERVICE

On 23 March, 2001, ground controllers in Russia relayed the commands that put an end to Mir's existence. Most of the station burned up in the Earth's atmosphere, but some 25 tonnes of Mir debris fell into the Pacific Ocean. Observers watched the flaming fragments leaving long comet trails in the sky.

point out that the results obtained were hardly proportionate to the exorbitant costs of maintaining the station, whose budget was substantially reduced following the dissolution of the Soviet Union in 1991. Mir only survived thanks to the co-operation of the USA and other countries, with a first American astronaut, Norman Thagard, taking up residence in 1995. Despite a string of problems, the station remained occupied until 1999. It was eventually brought down from orbit in 2001 (see box, left).

The International Space Station

Meanwhile, the USA and its collaborators had continued with work on the Freedom project, which came to seem more and more like a

Parting shot
A photograph of the International Space Station (ISS) taken from the space shuttle Discovery *shortly after completing a visit to the station in March 2009.*

larger version of Mir. By 1993 expenditure had soared to $30 billion. That year the Russians accepted an invitation from President Clinton to participate in the programme, whose name was subsequently changed to the International Space Station (ISS). The first module was launched in 1998, financed by the Americans but built by the Russians. A second, American-built module docked with it a few months later, and fresh habitation and maintenance modules were added in the years that followed. An initial three-person crew arrived in 2000, and the first space tourist, Dennis Tito, followed a year later, spending almost eight days in orbit.

Over the years the ISS has steadily grown. The building project is due for completion in 2011, by which time the station will be 108m (350ft) long by 74m (245ft) wide, weighing in at 415 tonnes and with a total living area of 1,200m³. The biggest contributor to the fabric of the station has been the USA, followed by the Russians, who have contributed about a third. Japan and the European Space Agency both have laboratories on board and the Canadians supplied a robotic arm. With the American Space Shuttle programme coming to the end of its operational life, the task of resupplying the station has increasingly fallen on Russian Soyuz manned spacecraft and Progress unmanned freighters. Costs have now risen to upwards of $150 billion for the 16 participating nations, and debate about the usefulness of the station is increasing. In all, more than 215 people have visited the station so far. The arrival of three new crew members in April 2011 brought the crew back up to six, currently five men and one woman, overseeing experiments that are for the most part automated. As with Mir, the hopes raised by the prospect of experimenting in zero-gravity conditions have not been fully met, as the new medicaments and materials once envisaged have failed to emerge and no fundamental breakthrough has been achieved in physics. One tangible gain has been in our understanding of the human body and the acceleration of the ageing process in conditions of weightlessness. On current plans, the ISS will continue to orbit the Earth 15 times a day until the year 2020.

A CHINESE SPACE STATION?

In 2009 China announced its intention to put three space station modules into orbit. The station is planned to be operational in 2020, with a crew of three on board.

FLOATING PLANTS

Most of the experiments carried out on the ISS have been conceived with an eventual manned mission to Mars in mind. On a trip of that duration the astronauts would need to grow much of their own food, so studying how plants grow in conditions of weightlessness has been a priority. Trials have shown that after a period of erratic growth caused by the absence of gravity, vegetables start to respond to other factors, such as the provision of light and water.

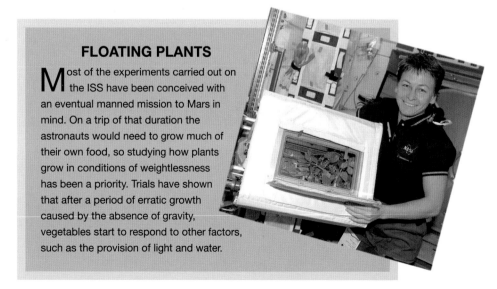

INVENTIONS THAT CHANGED OUR LIVES

The cash machine 1971

The automated teller machine, or ATM, is a telematic device (one linking telecommunications and computer technology) that allows bank customers to carry out transactions – in particular, withdraw cash – without requiring the services of a cashier. It was not a new idea. The first automatic cash machine was developed by American inventor Luther Simjian and installed in New York in 1939, but it was abandoned six months later for want of customers. In the late 1960s Scottish engineer John Shepherd-Barron took up the idea, devising a self-service cash dispenser operated by special cheques impregnated with a radioactive carbon-14 compound. Customers inserting the cheques had to identify themselves by entering four-digit personal identification numbers (PINs). But it was the patent taken out by American engineer Donald C Wetzel on 7 October, 1971, that brought about the ATM, or cash dispenser, so familiar today. His method combined a plastic bank card, incorporating a magnetic strip, with a PIN number, allowing users to access their own accounts and withdraw cash.

Dealing with a machine
Whether dispensing cash or sweets, automatic teller and vending machines have one thing in common: there is no-one to help if something goes wrong – as seems to be happening to this customer in the 2008 film Eagle Eye *(right). Confectionery vending machines made for the Austrian firm of Pez, like this 1965 model (left), have become collectors' items.*

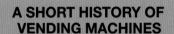

A SHORT HISTORY OF VENDING MACHINES

Automatic vending machines have a history stretching back into antiquity. In 215 BC the Greek mathematician Hero of Alexandria devised a fountain that provided water for sacrificial rites on receipt of a 5-drachma coin. The first patented systems appeared in the 18th and 19th centuries – a machine distributing postage stamps appeared in Britain and a sweet dispenser in the USA. But it was in the 20th century that the devices really caught on, thanks to the development of electronics and wireless technologies.

ATMs on every High Street

Wetzel's device took off commercially. By 2010 there were more than 700,000 cash machines worldwide, offering customers the possibility not just of withdrawing money but also of accessing instant bank statements, paying in cheques and transferring funds between accounts. Some are now operated by smart cards containing computer chips.

Soft contact lenses 1971

The first contact lenses, made of glass, were introduced towards the end of the 19th century. They covered much of the white of the eye as well as the cornea and were uncomfortable to wear. In the late 1930s lens manufacturers opened up new possibilities by using Plexiglas, a transparent plastic now better known by the Perspex brandname. After the Second World War a German, Heinrich Wöhlk, and an American, Kevin Tuohy, found a way of trimming the edges of faulty lenses, producing small plastic ones that fitted neatly over the iris. It quickly became apparent that lenses with a diameter of just 9mm or 10mm allowed extra oxygen to reach the cornea, making them more comfortable to wear. For the next two decades these were the only type of contact lens available, but they became increasingly popular, especially in image-conscious Hollywood where early adopters included future US President Ronald Reagan.

Toy story

In 1952 a Czech chemist, Otto Wichterle, discovered a hydrogel that became known as HEMA (short for hydroxyethyl methacrylate). This naturally supple polymer had a capacity for absorbing water, making it suitable for medical use. He saw its potential for contact lenses, but ran up against manufacturing difficulties, compounded by political troubles when his dissident views aroused the hostility of the country's Communist rulers.

Wichterle eventually cracked the problem of moulding the lenses when he realised that centrifugal force could be used to spread the gel as required. At home on Christmas night in 1961, he used his children's construction kit and a bicycle dynamo to improvise the equipment required, thereby producing the first soft contact lenses. He tested them successfully the very next day and applied for a patent on the process before the year was out. Sadly, Wichterle never saw the rewards of his discovery: the Czech authorities sold the patent rights without his knowledge to a US firm, Bausch & Lomb – a bizarre move in Cold War days.

To market

The new owners won official approval to put the lenses on the market in 1971, and they were soon being worn widely to correct for short-sightedness, astigmatism and blurred vision. In 1982 a new manufacturing process reduced the price to such an extent that it became feasible to change lenses every week; daily disposables followed in 1995. Since the late 1980s tinted versions and lenses with a protective covering against ultraviolet have also been available.

A RIGID ALTERNATIVE

Rigid lenses made of gas-permeable materials are easier to wear than the earlier hard lenses because they allow oxygen to reach the cornea. This even allows people with naturally dry eyes to use corrective lenses.

Construction-kit creation
Otto Wichterle used his children's construction set to make the first soft contact lenses. He lost his job as director of Prague's Institute of Macromolecular Chemistry in 1970 as a result of his support for democratic reform, but continued his research work until shortly before his death in 1998.

INVENTIONS THAT CHANGED OUR LIVES

The monoski and snowboard 1971

In the late 1960s adventure sports enthusiasts dreamed of re-creating the thrill of surfing on snow. An American, Mike Doyle, developed a board, made of glass fibre and resin, with two parallel forward-pointing bindings. He first tried out his monoski in 1969 and found it did not sink into powdery snow – and it hurtled down slopes at great pace.

Four years earlier Sherman Poppen had devised the snurfer, a wide board with a cord fixed to the tip for steering. The device was intended primarily for children until, in 1971, Jake Burton Carpenter had the idea of adding bindings. At roughly the same time Dimitri Milovich designed a board whose swallowtail design recalled surfboards, setting the bindings at an angle and marketing his invention as the Winterstick. Both men claimed the credit for inventing the snowboard, but the marque that survived was Burton's.

Snowboards and monoskis benefited in the 1970s from advances in skiing technology. These included the use of metal patterns for improved grip on hard snow and high-density polyethylene bottoms for a better glide.

Growing recognition

In the early days the sports were limited to a small coterie of enthusiasts who clambered up steep slopes at dawn to blaze trails in powdery snow. There was little direct competition, for monoskiers were concerned primarily with speed while snowboarders concentrated on turns. Little by little, snowboards and monoskis started turning up on recognised ski slopes, causing concerns for the resort authorities, who in the early days tended to ban boards for fear of accidents.

Over the years the boards became lighter, thanks to the adoption of a sandwich structure with a core of wood or polyurethane foam and surfaces made of epoxy resin reinforced with carbon fibre or glass fibre. The bindings also changed, becoming less like those on skis: in the 1980s step-in designs were introduced, making it possible to snowboard in soft shoes.

Meanwhile the first competitions were arranged, featuring downhill and slalom events as in skiing. In the 1990s new disciplines closer

Two styles of snowboarding *Freeriding (above) is usually done off piste and exponents must improvise their moves over whatever the terrain puts in front of them. Freestyle (left) is more spectacular, emphasising acrobatics performed over man-made obstacles.*

to skateboarding put in an appearance, notably freestyle, which featured acrobatic leaps and other tricks and was staged on mogul (humped) terrain for monoskis and in a half-pipe – a semi-cylindrical basin hollowed out of the snow – for snowboards. A whole culture developed around the two sports, featuring alternative dress styles and a laid-back approach recalling that of the water surfing community.

Skiing makes a comeback

In the late 1980s the competition from snowboarding was such that it caused a decline in the popularity of monoskis, which were the less manoeuvrable of the two except on powdery snow. Ski resorts reflected the trend by creating snow parks for the exclusive use of snowboarders and installing larger chairs on ski-lifts.

Another development of the time was a revival in the appetite for skiing. Taking their inspiration from snowboards, designers created skis that were parabolic or 'shaped', shorter than had previously been the norm and with a narrowed waist, making it easier to do carved turns without skidding. The new products worked as well on powdery snow as they did on moguls, and suited experts as well as beginners for their greater manoeuvrability.

Snowboarding became an Olympic sport in 1998, featuring Halfpipe and Giant Slalom. In 2006 a snowboard cross event was added, influenced by the sport of motocross. Some 1.7 million snowboards were sold worldwide in 2002 – evidence of the appeal that the activity has as a winter sport.

As for skiing – which had already turned acrobatic with the introduction of moguls, aerials and ballet in the freestyle demonstration at the 1988 Winter Olympics – it moved even closer to snowboarding with the introduction of a ski cross event at Vancouver, Canada, in 2010. It remains the world's leading winter sport.

SPORT FOR THE DISABLED

From the start of experimental devices in the 1970s, variations on the monoski have been developed to allow paraplegics to enjoy winter sports. By the early 1980s, Europeans were using 'ski-bobs' with chair-like attachments mounted on two small skis. The first Olympic medals for the sport were awarded at Innsbruck in 1988.

Downhill champion
Xavier Duret competing in a monoski event in the Hautes-Alpes region of France in 2005. At one time, the Duret family firm were producing 80 per cent of the world's monoskis.

NEW ARRIVALS

First seen in the late 1980s, snowkiting brings together snowboards and large kites to re-create something of the kitesurfing experience on the slopes. Skwaling, dating from the early 1990s, crosses skiing with snowboarding in that skwalers stand with their feet one behind the other on a single board, narrower than a snowboard and particularly well suited for carve turns.

THE PERSONAL COMPUTER – 1972
Information technology enters the home

At the start of the 1970s, a new generation of computers arrived that could fit onto a desktop. Aimed at individuals rather than big business, these machines would revolutionise life at work and at home.

In 1972 France's National Institute for Agronomic Research needed small computers. The machines available on the market at the time were cumbersome and costly, so the Institute asked the information technology (IT) firm R2E to develop something more suitable. The result was the Micral, launched in France in 1973, which quickly eclipsed a more basic device, the DIHEL Alphatronic, of a German competitor. Even so, both machines excited wonder at the time simply for the fact that they fitted on a desktop. In effect, the Micral, invented by François Gernelle, an immigrant from Vietnam, was the world's first real home or personal computer. In future IT would be within the reach of individuals, not just government departments and corporations.

Central unit and terminals
Students at Stanford University in California follow a Russian language course in 1967 (above). Their terminals were linked to a PDP-8 computer (right), which provided instructions for their learning exercises.

THE PERSONAL COMPUTER – 1972

PC pioneer
François Gernelle (2nd from left) with colleagues and a Micral in 1977. The computer was equipped with an Intel 8008 microprocessor and was inexpensive by the standards of the day.

Small is beautiful

The quest for computers that were smaller and easier to use than mainframes had got under way early in the previous decade with the development of minicomputers. In 1961 a US firm, Digital Equipment Corporation, introduced the PDP-1 (Programmed Data Processor), the first computer that was affordable for small companies or laboratories. It was the first interactive computer: the user could communicate with it using a keyboard and a screen. In 1964 the same firm brought out the PDP-8, which was even more compact, cheaper and easier for beginners to use thanks to the Focal computer language. The device was soon a favourite in research laboratories.

The big boys arrive

A new generation of entrepreneurs now entered the scene. In the USA Bill Gates and Paul Allen developed the operating system and software for the Altair 8800, the first personal computer mass-produced in kit form for individual users (although without a keyboard or screen), which was introduced in 1975. Soon after they founded Microsoft. In the same year the IBM 5100 was launched, described as an 'all-in-one' personal computer – this time complete with keyboard and screen – that simply needed plugging in. Yet few units sold because it was slow and the screen was inadequate.

In 1976 Steve Jobs and Steve Wozniak assembled a computer in kit form in a

Do-it-yourself kit
This computer (below), designed by Bob Belleville in the 1970s, came in kit form. The hard disc, screen, loudspeaker and keyboard all came packed in a wooden box. It was equipped with an Intel 8080 microprocessor.

DO YOU SPEAK ASCII?

Computers are designed primarily to handle numbers. To cope with text, the letters, symbols and punctuation marks all have to be given a numerical coding. That was the job of ASCII (the American Standard Code for Information Interchange), created in 1965 by Bob Berner, a US citizen. ASCII quantified 128 characters, but had no codes for accents. These had to be added when computing went international, but they initially differed between the various languages using the system, which complicated the business of exchanging files. To solve the problem Unicode was introduced as a new universal standard, but it proved cumbersome – a piece of text in Unicode takes up twice as much space as one in ASCII. As a result some systems do not use it, and the UTF-8 code has been introduced to address its shortcomings.

The computer revolution in action
A young woman learns to use a computer as part of a jobseekers' training programme in London in 1975 (below). By the 1990s computer skills were widely seen as essential requirements.

Californian garage and the Apple I was born. Their ambition was to take computing to the man and woman in the street. Although it had a keyboard and screen, their first attempt was at best rough and ready and it failed to sell. They had more luck the following year with the Apple II, which came ready assembled, equipped not just with keyboard and screen but also with an audio-cassette interface for feeding in data. The model soon saw off the TRS-80, manufactured by Tandy Radio Shack, and the Commodore PET, both of which came out that same year. The Apple II quickly sold more than 2 million units, making it the standard-model home computer at the time.

The apple core
The Apple I (above) comprised a single circuit board, a colour screen and a keyboard.

The coming of the PC

In 1981 IBM, which had initially steered clear of the home computing market, launched its own personal computer. The IBM PC used an operating system provided by Microsoft, which was also supplied to manufacturers of 'clones' – computers resembling the IBM machine but assembled by rivals – which multiplied from 1983 on. One result was to entrench the IBM system as the industry standard, forcing even proficient competitors like Commodore out of the running, except for specialised multimedia niches. Meanwhile, Microsoft established its dominance over operating systems and the associated software. Intel, which produced the chips for the IBM PC and compatible machines, became the biggest microprocessor manufacturer on the planet. Apple alone continued to resist the tide, surviving thanks to enduring innovations, particularly in design and the human/machine interface.

THE SINCLAIR ZX81

The machine that did most to introduce home computing to the UK was a small and light device – the Sinclair ZX81. Introduced in 1981, it was stripped to the bare essentials to keep costs down, and had to be used in conjunction with a cassette player (to load programmes) and a television set (to provide a screen). It sold for less than £50 in kit form and proved a huge success, selling over 1.5 million units. Other popular early British models were the BBC Micro and the Amstrad PCWs.

Apple innovation
An important step in making computers more accessible was the arrival of the Macintosh Lisa which introduced such user-friendly features as scrolling menus, a waste bin, copy-and-paste instructions, and a mouse complete with a button.

COMPUTER LANGUAGES AND OPERATING SYSTEMS

Instructions have to be fed to computers in sequences of the two digits '0' and '1', the only 'language' they can understand. The trouble is that humans are not used to thinking in that way. To write computer instructions, programmers use an intermediary language that is then translated into 0s and 1s by a special programme known as a compiler.

Programming languages

The first programming languages were called assembly languages and were still close to machine-speak. In 1957 IBM introduced Fortran, the first universal language, which was joined by BASIC in 1964. There are now dozens of languages, some of them highly specialised.

Object-oriented programming (OOP) first appeared in the early 1970s. Before that, all the basic operations had to be spelled out one after the other in the order in which they were to be executed. With OOP, a computer was given instructions to apply certain actions to objects (whether physical entities, ideas or concepts), as for example the command 'Advance the knight' might be given in a game of chess. In classic programming, applying a similar action to a different object – for example, giving the command 'Advance the pawn' – involved rewriting all the instructions from scratch. But OOP defined each object once and for all, including whatever was particular to its behaviour and its interaction with other objects (in the example given, for instance, a knight does not move forward in the same way as a pawn). Thereafter there was no need to know how exactly an object was programmed in order to write another programme involving it – say, 'Move the knight back' – for the basic actions themselves were pre-programmed. OOP considerably speeded up the business of writing software. It presupposed that everyone was describing objects and actions in the same way, a process of unification that is being carried forward today by the Unified Modelling Language.

Operating systems

A computer's operating system is an indispensable implanted programme that controls the processor, prioritising the various software programmes involved, as well as the operation of RAM memory (see box, page 106) and any attached peripherals, such as printers and scanners. It guarantees the machine's security, monitoring access to files and applications, and supervises the proper functioning of software programmes. It also carries out commands given by the computer user.

The first operating system was created in 1954 by Gene Amdahl, a Norwegian-American working for IBM. Among the most influential have been Unix (1970), CP/M (1973) which was the first system designed for microcomputers, and QDOS, later renamed PC-DOS because it was used in IBM PCs before being bought back by Bill Gates and marketed as MS-DOS. Unix, Windows, GNU/Linux and the Macintosh operating system dominate the market.

Computerspeak – a foreign language
Programming languages can attain a degree of complexity barely guessed at by most computer users, who experience only the user-friendly interfaces.

Computer bird
Tux the Penguin (above) is the icon of the Linux operating system, created in 1991 by Linus Torvalds of Finland and employing mostly free software.

THE PERSONAL COMPUTER – 1972

RAM AND ROM

RAM (Random Access Memory) is the computer's working memory. Containing the data currently being handled by the processor at any given moment, it is modifiable and operates fast, but is also volatile – data is lost if the current cuts out. RAM capacity determines the number of bytes of information that the computer can handle. ROM (Read Only Memory) provides permanent storage for information, preserving it even when the machine is switched off, but it cannot modify that information. ROM contains the unvarying parts of the system, like the BIOS (Basic Input/Output System), which 'wakes up' the computer when it is first turned on, verifying that everything is working properly and launching the operating system.

The Macintosh Classic
Marketed from 1990 to 1992, the Classic had 1 megabyte of RAM and 512 kilobytes of ROM.

ACRONYM EXPLAINED
The familiar IT term DOS stands for 'Disk Operating System' – the programme that controls all of a computer's operations.

Child's play
Children explore a Tandy Radio Shack 80, one of the first desktop computers that went on the market in 1977.

Desktops get user-friendly

Apple launched the Macintosh Lisa, supposedly named after Steve Jobs' daughter, on 19 January, 1983. It was the first personal computer to benefit from a graphical user interface (GUI). Previously people had only been able to communicate with their machines via the keyboard, typing in commands that appeared as lines of text on the screen. With the Lisa they encountered windows, icons and scroll-down menus that were accessed via a mouse. The arrangement was intuitive to use and enabled multi-tasking. In fact, the Lisa took its inspiration from work done at Xerox, where the first computers with these innovations had been produced – but were too expensive for the mass market.

In 1984 Apple followed up with their game-changing Macintosh, named after a variety of apple. This used the same mouse, operating system and GUI features as the Lisa, but was much cheaper to buy and faster to use, being more focused in its applications. The Mac was a success, spawning a family of models that continues to this day.

A sociological phenomenon

Although they were not very powerful in the early days, home computers made rapid progress thanks to ever-faster processors and steadily enlarging memories. Meanwhile, the audio cassette inputs gave way to floppy disk drives, then to CDs and DVDs. The original tiny screens – in monochrome green for IBM PCs and in eight colours for the Apple II – grew bigger, bursting into full colour in the mid 1980s. Today's personal computers are powerful enough to reproduce moving images with a high degree of clarity. As a result information technology ceased to be the exclusive preserve of large organisations and research laboratories. Shops, small businesses and the liberal professions all took it up. In most workplaces each employee was supplied with a computer, often linked to others through an internal network. No longer adapted

Control room
Computer screens in the control room of the Civaux nuclear power plant in France (above) enable the plant's operatives to monitor and regulate the plant's output.

Design job
Produced from 1998 to 2003, the iMac G3 (left) was the first 'designer' computer, styled by the British designer Jonathan Ive. Initially released in Bondi blue, it eventually came in 13 colours and patterns.

THE LINUX KERNEL

On 5 October, 1991, Linus Torvalds, a Finnish student, posted the first version of an operating system that he had developed on Usenet – an on-line discussion forum that predated the Internet. He wanted to construct a multi-tasking, multi-user system better than the one previously installed in his computer. What he came up with was Linux. Thousands of well-disposed programmers made use of it, suggesting ways in which it could be improved – as they continue to do to this day. Linux is a kernel, the heart of an operating system, controlling a computer's memory and the software feeding into the microprocessor, as well as access to peripherals (such as printers and scanners) and to the network. An array of free software known collectively as GNU (said to be short for 'Gnu is not Unix') completes the system, which is described as 'open' in that anyone can distribute it and can access and alter the underlying source code. Many different distributions – packages of software applications for which there is usually no charge – are now available to run on Linux.

primarily for scientific calculations, the new desktop computers were better suited to creative work or administrative tasks thanks to software specifically designed to handle office and accounting functions and images. Like every new technology, the spread of personal computers has cost some jobs, but it has also created new ones, notably in the thriving software industry and computer support.

In recent years prices have fallen so low that computers have found their way into school classrooms and people's homes. They have become a part of daily life, whether for work, practical requirements like shopping, or home entertainment. Together with the Internet, they have created the information and communication society. Today a billion home computers are in use around the world, most of them in the developed lands, where half of all households are thought to possess them.

THE DISPOSABLE SOCIETY
Buy it and bin it

In the 1960s and 70s re-usable dish cloths, glass bottles and fountain pens rapidly lost ground to paper towels, plastic containers and ballpoints, among other ephemera. Consumers had entered the disposable era, a corollary of a society linked to economic growth.

Some disposable inventions long predated the latter part of the 20th century, such as the non-rechargeable battery which hales from 1886. Many first put in an appearance in the culinary and household fields, responding to changing requirements in matters of hygiene and ease of use. For example, in 1908 a German woman called Melitta Bentz grew tired of drinking coffee grounds and had the idea of putting a sheet of absorbent paper inside the coffee pot to strain the brew. The coffee filter was born and the brand still bears her name.

The First World War boosted demand for bandages and accelerated research on cellulose, which was fundamental to an invention that would revolutionise women's lives: the sanitary towel. Created for Kimberly-Clark in 1920, it was followed four years later by Kleenex paper hankies, produced by the same firm. A US doctor, Earle Haas, invented the tampon in 1931, the same year that Scott Paper marketed the first paper towels. Initially launched in the USA, these novelties reached Europe after 1945, where they proved something of a stimulus to innovation. Disposable nappies were invented in Britain in 1948, but it was a US brand that cornered the market when disposables really began to take over from washable nappies in the early 1970s.

The age of plastic

The throw-away revolution made its way into almost every field of life with the growing use of plastics. Invented before the Second World War, these could now be produced cheaply and in huge quantities thanks to cheap crude oil. The early 1950s saw the arrival of disposable ballpoints, invented by the Hungarian brothers Lazlo and Georg Biro (see box, opposite), and cigarette lighters, ushering in a new era stimulated by improvements in production processes and by fresh distribution circuits.

Nappie mountain
An Environment Agency study in 2005 estimated that 400,000 tonnes of disposable nappies are thrown out in the UK annually.

Supermarkets were by then spreading around the world, seeking to sell ever-increasing quantities of rapidly restocked products. Partly in consequence, food packaging became lighter, cheaper and more convenient. The ring-pull aluminium can was invented by US equipment manufacturer Erman Fraze in 1959. In 1941 two English chemists, Rex Whinfield and James Dickson, working in Manchester at the Calico Printers' Association, had developed polyethylene terephthalate (PET or PETE). By the 1960s this was becoming popular for packaging soft drinks, due to its strength, lightness and impermeability – though it took until 1994 for the Coca-Cola

BIC – A CASE HISTORY

In the early 1950s Marcel Bich, the head of a firm manufacturing parts for pens, bought up the patent on the ballpoint pen taken out in 1938 by the Hungarian, Lazlo Biro. Bich marketed the pens under the Bic trademark, dropping the 'h' from his name for easier pronunciation. In 1952 he launched the Bic Crystal, a cheap ballpoint designed to be used then thrown away. In 1973 he moved into the market for disposable lighters (right), invented by Jean Inglessi in 1948, making them non-refillable. Two years later he swelled his fortune by launching disposable razors. Bich, who loved the sea, shifted his attention for a while to making sailboards, but soon returned to disposable products. One of his few failures was the disposable perfume spray, sold for small change from tobacco kiosks. Bich died in 1994 but his heirs continued along the same path, teaming up with Orange in 2008 to launch the disposable Bic Phone, another success. That same year the firm took a step towards sustainable development with pens made from recycled plastic and razors produced from a bioplastic derived from corn starch and coloured with natural dyes.

Short shelf lives *Cardboard and plastic disposable plates and glasses have now been joined by more biodegradable products derived from palm, bamboo, wood or corn starch.*

corporation to make the transition from glass. At about that time too, traditional double-handed shopping bags started to lose ground to plastic alternatives.

All these products suited the new lifestyles of the day. In the 1960s throw-away paper plates and plastic glasses became a familiar part of picnics and holidays. Disposable razors, invented in 1975, allowed men to save time by no longer needing to change blades and clean their shaving equipment.

Upping the ante

By the early 1970s the consumer revolution was in full swing. Electrical appliances, hi-fis and TV sets, many of them made in Japan, were becoming steadily cheaper as production costs were pared back. If a device broke, getting it fixed came to seem a bother, with spare parts often difficult to obtain. 'Replace, don't repair' was the new watchword, dictated by the tenor of the times.

A whole range of downmarket disposable gadgets – such as calculators and watches whose batteries could not be replaced – now came on the scene, manufactured at rock-bottom rates in low-cost countries. The trend extended to relatively hi-tech developments: throw-away contact lenses were introduced in 1982; the disposable camera, Fuji's Quicksnap, first appeared four years later, made of plastic with a film already loaded. Versions soon also became available with a built-in flash and a degree of waterproofing.

Ready for reuse *The aluminium used to make cans is endlessly reusable. Over 30 per cent of the aluminium employed in developed countries comes from recycling.*

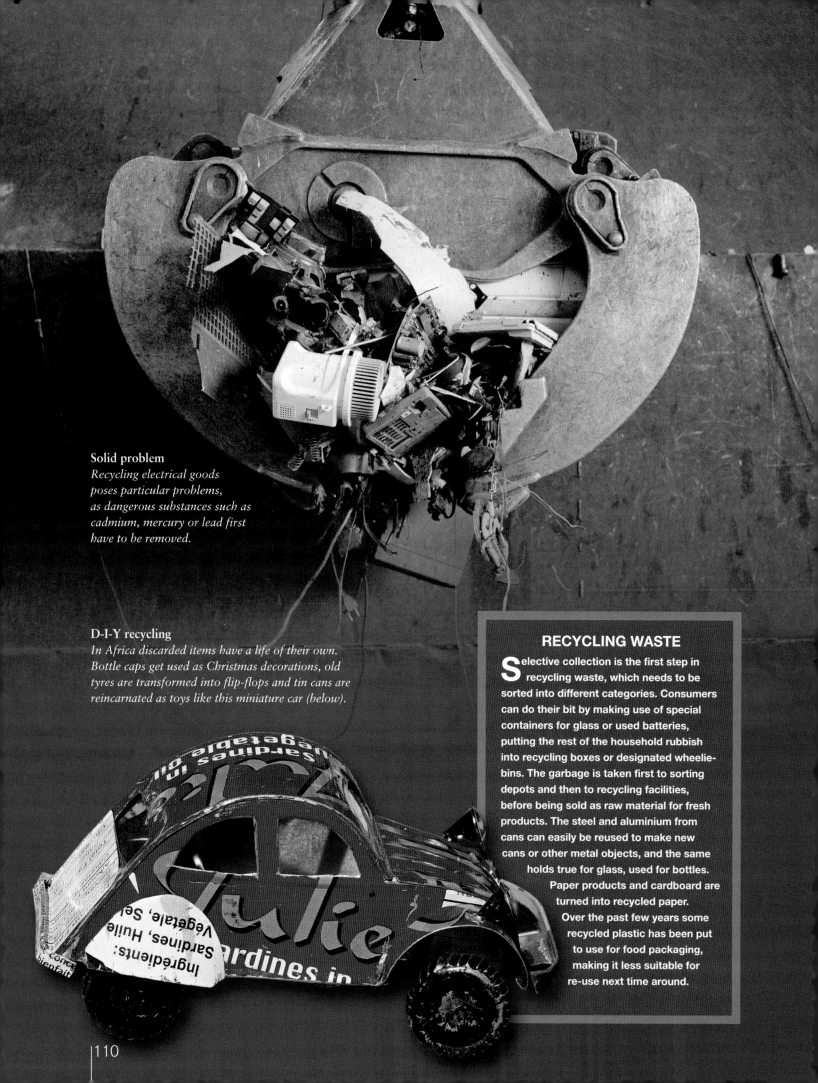

Solid problem
Recycling electrical goods poses particular problems, as dangerous substances such as cadmium, mercury or lead first have to be removed.

D-I-Y recycling
In Africa discarded items have a life of their own. Bottle caps get used as Christmas decorations, old tyres are transformed into flip-flops and tin cans are reincarnated as toys like this miniature car (below).

RECYCLING WASTE

Selective collection is the first step in recycling waste, which needs to be sorted into different categories. Consumers can do their bit by making use of special containers for glass or used batteries, putting the rest of the household rubbish into recycling boxes or designated wheelie-bins. The garbage is taken first to sorting depots and then to recycling facilities, before being sold as raw material for fresh products. The steel and aluminium from cans can easily be reused to make new cans or other metal objects, and the same holds true for glass, used for bottles. Paper products and cardboard are turned into recycled paper. Over the past few years some recycled plastic has been put to use for food packaging, making it less suitable for re-use next time around.

THE DISPOSABLE SOCIETY

D-I-Y HOME DÉCOR

In the 1970s there was something of a revolution in the home furnishing business. Following the general market trend of the day, furniture became steadily cheaper, allowing householders to change the décor in their homes as fashion and their own taste dictated. This golden age of plywood and formica faltered in the 1990s as ecological awareness spread, along with a desire for a more sustainable form of consumerism. A new drive towards do-it-yourself decoration now became evident, with people seeking to ring the changes with what they already possessed. An old chest of drawers might be stripped down and painted, say, or kitchen cupboard doors replaced instead of buying whole new units. In the first decade of the new century a flood of books, magazines and television programmes provided tips for Sunday decorators and budding interior designers. Customisation became the new alternative to consumption.

Plastic all around
Plastic goods come in thousands of different forms, from disposable cameras (above) to mini-calculators like this one on a key-ring (left), designed for currency conversion. Some plastics are also increasingly biodegradable: the pen below, for example, is made from a bioplastic derived from corn starch.

Reversing the trend

There was a downside to this new-found convenience. By the early 1990s the world's rubbish bins were overflowing with jumbo-size packing cases, and land-fill sites were fast filling up. Discarded washing machines, TV sets and fridges were becoming unwanted man-made mountains. By the year 2005 it was estimated that the average British household was throwing out half a tonne of waste per person annually. Packaging made up a significant part of the total, as did sanitary products such as wipes, paper handkerchiefs and especially disposable nappies.

By this time, attitudes were starting to change. Non-disposable products appearing on the market included washable nappies, menstrual cups and new-generation rechargeable batteries. Far from harking back to the old days, these environmentally friendlier goods were marketed at modern lifestyles. At the same time, often in response to consumers, manufacturers cut back on packaging and began to encourage recycling. More importantly, disposables themselves were being transformed, as in recycled paper products including toilet paper, kitchen rolls and paper tissues. Even plastic, once the non-biodegradable material second to none, has begun to be recycled or else replaced; potato starch, for example, is used to make disposable bags. And many supermarkets have stopped automatically handing out free plastic bags.

Today the battle-cry is 'Consume less, consume better'. A new era of safeguarding resources might just be dawning.

THE POCKET CALCULATOR – 1972
Higher maths in the palm of your hand

An end-product reflecting remarkable technological advances, the pocket calculator marked the start of something of a revolution. It was soon taken for granted, not just in scientific circles but also in daily life.

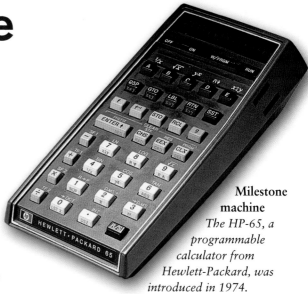

Milestone machine
The HP-65, a programmable calculator from Hewlett-Packard, was introduced in 1974.

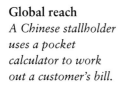

Global reach
A Chinese stallholder uses a pocket calculator to work out a customer's bill.

The history of calculators reflects the wider story of computers as a whole. In the mid 20th century, they were cumbersome machines that were heavy and difficult to move around. Thanks to the invention of transistors, they shrank in size, but they were still only used by businesses. The introduction of integrated circuits enabled engineers at the US electronics firm Texas Instruments, notably Jack Kilby, to produce a prototype pocket calculator in 1967, but the cost of the chips that went into the machine meant that it remained expensive. Then the Apollo programme, which used millions of integrated circuits, helped to bring the price down. Texas Instruments took the opportunity to refine their original model and put their TI2500 model on the market in 1972.

From cutting-edge tool …

In the rush to innovate, Hewlett Packard had no intention of leaving the field free for their competitors at Texas Instruments, and the same held true for the Japanese firms Sharp, Casio and Brother. Improvements in the display, the power supply and the layout of the keyboard produced more sophisticated models. The first calculators had only been able to add, subtract, divide and multiply, but devices now appeared that could work out square roots and had trigonometric functions. Hewlett Packard built a reputation for

---

REVERSE POLISH NOTATION

Reverse Polish notation (RPN) is an easier way of writing arithmetical formulae. Instead of entering 2+2 on a calculator, for instance, it involves keying in 2 2 +. Derived from a system developed in 1920 by the Polish mathematician Jan Lukasiewicz, it follows a logic that works for all types of calculation, by which the user first provides the figures, then indicates what is to be done with them. In practice, RPN saves time by reducing the number of keys that need to be pressed. Mainly used for scientific purposes, this calculating method was employed, among other uses, to facilitate the Apollo 11 Moon landing.

CALCULATORS IN THE CLASSROOM

The use of calculators in schools remains controversial, with their backers claiming that they spare children the necessity of doing boring sums while critics maintain that they prevent them from acquiring the habit of mental arithmetic. Cost was initially an obstacle to their adoption, but that is no longer the case, and they are now standard equipment from primary years up to 'A' levels and beyond. By 1994, 87 per cent of primary and 50 per cent of secondary schools in England were using them regularly. Yet no real attempt has been made to assess statistically the role they play in learning. Although they are now used almost universally in maths classes, there is still debate about how best to employ them as tools for exploring the world of numbers.

scientific calculators with its programmable HP-65 model, employing reverse Polish notation (see box, left), the calculator selected to accompany the Apollo-Soyuz astronauts on the 1975 joint test project. By that time competition was heating up with the appearance first of other programmable devices, then of graphing calculators.

... to everyday convenience

Since that time the market in calculators has split in two. On the one hand the scientific and financial sectors have fuelled demand for powerful calculating machines in compact formats. On the other, cheap and easy-to-use models have flooded the planet, becoming everyday devices used by all and sundry in homes, offices and schools. Current estimates suggest that more than 10 billion pocket calculators have been sold over the past four decades. Mass-produced in China and Southeast Asia, calculators have lost their technological mystique so completely that they have come to be regarded by some as little more than useful free gifts for businesses to give away to customers. Tourists carry them to convert currencies, and they are accessorised as extra features on mobile phones or computers.

There is a downside. With the spread of pocket calculators, people have increasingly lost the habit of performing mental arithmetic, a skill that is fast disappearing as people use a calculator for even the simplest of sums.

Sun power
Many calculators today, like the one above, are solar-powered, equipped with photovoltaic cells that convert natural or artificial light into electrical energy.

CHAOS THEORY – 1972
Edward Lorenz and the butterfly effect

Chosen by the conference organisers against Lorenz's wishes, the title of a lecture that the US mathematician and meteorologist gave on 29 December, 1972, was deliberately provocative: *Predictability: Does the Flap of a Butterfly's Wings in Brazil Set Off a Tornado in Texas?* The phrase circled the globe, marking the birth of deterministic chaos theory.

The lecture marked the culmination of Lorenz's work on chaos theory, which had held his attention for almost a decade. In 1963, while working at the Massachusetts Institute of Technology (MIT), he had constructed a climate model for purposes of computer simulation. Although he managed to translate the dynamics of the global weather system into relatively straightforward equations, Lorenz soon discovered to his surprise that the results he obtained changed radically if he so much as altered by a few thousandths the numbers that he entered into the computer.

A blow to determinism
The realisation came as a shock. Lorenz understood that the extreme sensitivity of the model touched on a fundamental tenet of physics: the belief, held since the 19th century, that if known equations could describe the workings of a system, then it should also be possible to predict likely outcomes. He revealed his findings in an article entitled 'Deterministic non-periodic flow', limiting them for the time being to meteorology. Despite his caution, chaos theory gradually acquired a life of its own, due in part to the work of French physicist David Ruelle and Floris Takens of the Netherlands, who in 1971 extended the principle to astronomy, mechanics, chemistry, biology, ecology, electronics, optics, fluid dynamics and other fields. Another key contributor was the French-American Benoit Mandelbrot: using a home computer, he pioneered the mathematics of fractals (a term he coined in 1975), which helped to describe and picture the actions of chaos, rather than explain them.

Small cause, big effects
Addressing the issue raised in the title of his ground-breaking lecture, Lorenz stated: 'Lest I appear frivolous in even posing the question, let me try to place it in proper perspective by offering two propositions. (1) If a single flap of a butterfly's wings can be instrumental in generating a tornado, so also can all the previous and subsequent flaps of its wings ... and (2) If the flap of a butterfly's wings can be instrumental in generating a tornado, it can equally well be instrumental in preventing a tornado.'

A three-body problem
Imagine, for example, an astronomical system composed of three separate stars. If their mass, position and velocity at any given time are known, the general laws of gravitation should in theory be able to perfectly describe the future trajectory of each one. But in practice, no physicist can predict where those stars will be in a few million years' time, no matter how

---

A FRUITFUL MISTAKE
Lorenz stumbled on deterministic chaos theory through an error. In 1963 he was testing climate models by feeding them through a computer. Protocol demanded that he carry out the operation twice for each simulation, using the same parameters, which were entered manually. In the course of one of these double tests he mistakenly altered the data by a factor of one-thousandth, and by doing so obtained a totally different result. His scientific mind rapidly grasped the fact that, because of the error, he had put his finger on a fundamental phenomenon of physics.

powerful and computationally perfect the super-computer making the calculations. The system, then, can be described as 'chaotic'.

Strange attractors

All chaotic systems have one trait in common: a microscopic variation in the initial conditions (the flap of the butterfly's wings) provokes unforeseeable consequences (the tornado). Given that no measurement is exact – neither the position nor the mass nor the velocity of a body can ever be known with absolute accuracy – the evolution of the system in the long term is intrinsically unpredictable, not because of its complexity but by its very nature. The reverse of the coin is that any chaotic system has a peculiarity in that, however unpredictable it might be, it generally remains confined within recognised bounds. For instance, no-one knows exactly where the planets of the Solar System will end up at some distant future time, but it is possible to know that they will still be orbiting the Sun. This phenomenon is known as a 'strange attractor': it has an easily recognisable geometric structure and demonstrates the evolutionary boundaries of a system. Thanks to strange attractors, it is possible a priori to recognise physical systems whose evolution is chaotic – the attractors' main use. Lorenz modelled one in 1963, and the resulting graph looked surprisingly like … a butterfly's wing.

Confined chaos
Shown here in an artist's impression, strange attractors enable scientists to study seemingly chaotic phenomena that remain confined within identifiable limits.

INVENTIONS THAT CHANGED OUR LIVES

CT scanning 1972

A MEASURE OF DENSITY

The Hounsfield scale measures the density of body organs, rising from 0 HU (Hounsfield units) for water to 4,000 HU or more for bony tissue.

In 1963 a South African-born nuclear physicist by the name of Allan Cormack had the idea of upgrading the use of x-rays in medical imaging. As the rays traverse the organism, they are to a greater or lesser extent absorbed by the tissues, making it possible to obtain radiographs (x-ray images). Yet two-dimensional x-ray photographs are of only limited use, for the different structures pictured overlap on the image. Cormack believed that it should be possible to find a mathematical formula to describe the extent to which the rays are blocked along the length of their course and so to measure their depth.

Layering images

From 1967 on Godfrey Hounsfield, an English electrical engineer employed by EMI, sought to put Cormack's idea into practice. His goal was to reconstruct three-dimensional images of the body layer by layer, using a computer to analyse the blocking of the x-rays on different trajectories – a process known as x-ray computed tomography (CT). The machine he envisaged involved a moving beam scanning each layer of successive sections of the body, with detectors placed around it to capture the x-rays as they completed their passage. Having experimented on carcasses of cattle and pigs, Hounsfield collaborated with Jamie Ambrose, a radiographer at the Atkinson Morley Hospital in Wimbledon, to build a prototype that could be used on human patients. They got their reward late in 1972, when the machine won general approval for its sensitivity and precision at a medical conference in Chicago.

Initially used exclusively for brain scans, the device was soon being employed to image the entire human body. In 1979 Cormack and Hounsfield were jointly awarded the Nobel prize for medicine, with the citation lauding the diagnostic prospects opened up by scanners, which could reveal hitherto unnoticed anomalies including haemorrhages, infections, inflammatory lesions, tumours and cysts. Over the years scanning times have speeded up and the level of detail has increased, partly thanks to the use of injections of iodine-based radiocontrast agents to sharpen the image. CT scanning has become a routine medical procedure, although the radiation involved means that it is not entirely risk-free.

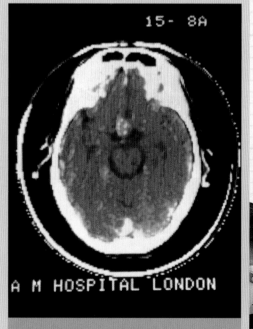

Medical breakthrough
The scanner at the Atkinson Morley Hospital in Wimbledon (below) produced the first brain scans, revealing in this case a patch of calcification (left, in red).

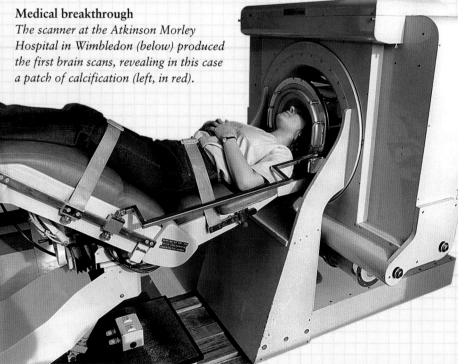

ALL-SEEING EYE

Because of the amount of detail they provide on the body parts they survey, CT scanners are used to check brains, eyes, vertebral columns, hearts, blood vessels, lungs and genitals, along with the digestive and other organs.

MRI scanning
1973

In 1969 a New York doctor, Raymond Damadian, decided to apply nuclear magnetic resonance (NMR) techniques to the study of organic tissue. NMR exploits the signals emitted by atomic nuclei in a magnetic field (see box below). By 1971, in experiments on rats, he was able to distinguish between healthy tissue and cancerous tissue from the water content of their bodies.

Paul Lauterbur, a chemist who attended some of the experiments designed to test Damadian's results, had the idea of capturing images of anatomical structures by submitting the body to a graduated magnetic field. As in a CT scanner, a three-dimensional picture of the tissues would be built up by mathematically analysing data obtained from sectional images. Lauterbur obtained his first images of small objects in 1973, and MRI – magnetic resonance imaging – was born.

Non-invasive examination

Peter Mansfield of Nottingham University refined the mathematical analysis of the radio signals involved, and in 1976 Damadian built a prototype device that could be used on humans. The first images, of a head and a thorax, were produced in 1979 and 1980, and hospitals began to install the equipment from 1982 on. Painless and non-invasive, an MRI examination is carried out in a metal cylinder known as the bore, big enough to completely surround the patient stretched out on a sliding table.

One major advantage of MRI over CT scanning is that it does not involve ionising radiation; another is that for many conditions it provides better images, making it the first choice for scanning the brain and other soft tissues and for diagnosing tumours, slipped discs and ligament injuries. Since 1991 functional MRI (FMRI) has been used to study the activity of different areas of the brain and to guide real-time surgery, particularly in the field of neurosurgery. By 2003, when Lauterbur and Mansfield received the Nobel prize for medicine, more than 60 million MRI scans were being conducted around the world each year.

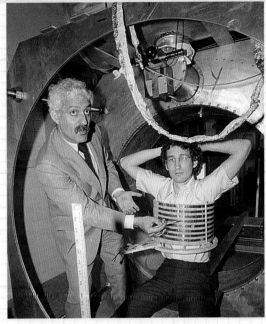

MRI prototype
Dr Damadian demonstrates the first practical MRI scanning machine in 1977 (left).

State of the art diagnostic tool
The MRI scanner at La Timone hospital in Marseilles is one of the most up-to-date in Europe (bottom). The 35-year-old patient shown being scanned was suffering from a slipped disc (below). The cost of such machines, at well over £1 million, has to some extent limited their use.

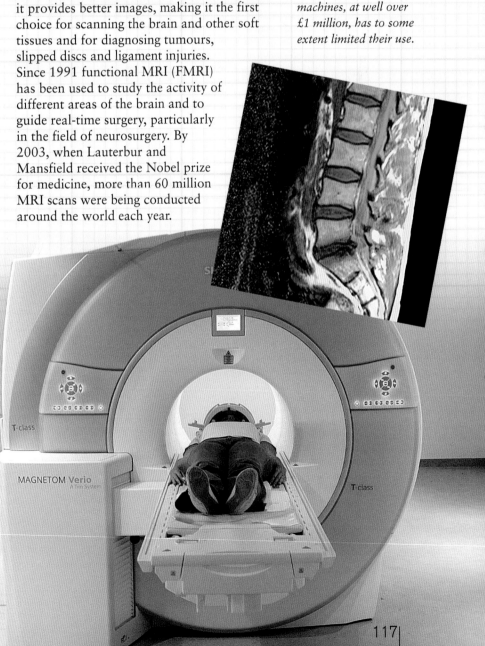

MAGNETISING THE NUCLEI

The NMR effect was discovered in 1946 by Felix Bloch and Edward Mills Purcell, winners of the 1952 Nobel prize for physics. In an intense magnetic field, the nuclei of hydrogen atoms in the water contained within tissues orient themselves like tiny magnets. When radio waves pass through them, some briefly change orientation and then return to their earlier state, emitting a signal on the same frequency as they do so (the resonance).

INVENTIONS THAT CHANGED OUR LIVES

The hard disk drive
1973

In the early 1950s IBM commissioned Minnesota-born Reynold Johnson to solve a crucial problem: getting rid of the bottleneck that built up when information was permanently stored in computers. The US engineer first explored the possibilities of a rotating drum containing a fast-moving magnetic head that could read data from or to tracks inscribed on its surface. The result, in 1956, was the Ramac 305, the first hard disk drive, although it could perhaps be more accurately described as the first data storage cupboard: weighing 1 tonne, it contained fifty 60mm disks serviced by a single head and could handle 5MB of information.

Progress thereafter was steady until, in 1973, IBM brought out the 3340. This had two separate modules, one permanent and the other removable, each with a storage capacity of 30MB. Improvements followed, but the basic principle had been established: the storage disks rotated rapidly around a central axis, and read/write heads moved over the disks generating a magnetic field that enabled them to input or access data. The heads polarised tiny areas on the surface of the disks that either did or did not trigger a current corresponding to digital '0's and '1's.

Expanding memories

The main thrust of progress since that time has been in storage capacity and the speed with which data can be accessed. In 1982 audio CDs signalled the arrival of laser technology and all the various incarnations of optical disks. Hard disk drives remained magnetic, but benefited from ever-expanding storage space and steadily shrinking prices. Today hard drives have an average capacity of about 160GB and are a central part of society's general trend toward increased amounts of both visual and textual information.

Happy birthday Ramac 305
The Ramac 305 photographed in 2006 (above), half a century after its launch. Between 1956 and 1961, IBM sold 1,000 machines at a cost of $10,000 each.

Miniaturisation in progress
In the year 2000 IBM introduced a hard drive barely bigger than a coin (left) with 1GB capacity. The 4MB Casio memory card (below right) came out in 1998.

MICRODRIVES AND FLASH MEMORY

In 1999 IMB introduced the first microdrive, a hard disk that could store 340MB of data in a device barely bigger than a box of matches. The development of flash memory – a small-format version of the random-access memory (RAM) found in computers – took the process even further. Flash cards are now common in devices ranging from telephones to MP3 players and USB drives.

The skateboard 1973

For some the natural successor to soapbox racing, for others a land-based extension of surfing, skateboarding had obscure origins. The first commercial skateboards were marketed in 1958 by two American brothers, Bill and Marc Richard, who made them by attaching roller-skate wheels to wooden boards and sold them through Californian surfer magazines. The sport only really took off after 1973, when Frank Nasworthy introduced polyurethane wheels, which were tough and had good roadholding qualities, producing boards that were faster, quieter and more stable. Today almost 20 million people use skateboards, performing feats of acrobatics in purpose-built skateparks or even on city streets.

CELEBRITY SKATEBOARDER

Born in California in 1968, Tony Hawk built his reputation in the 1980s and 90s, winning virtually every major skateboarding competition over the course of his career. He was the No.1 'vert' ('vertical') skater by the time he was 20, inventing 50 or so acrobatic tricks including the 900, a two-and-a-half turn aerial spin. He has since lent his name to a well-known series of video games.

Anti-lock braking 1973

In 1973 the German company Bosch got together with a firm called Teldix to work on an anti-lock braking system (ABS). The result was the *Antiblockiersystem*, patented and put on the market in November 1978, which did much to improve road safety over the ensuing years. Installed in cars, lorries and aeroplanes, it works on all four wheels at once, making it easier to control vehicles on wet or icy surfaces by improving directional stability and reducing the risk of skids.

ABS prevents wheels from locking by constantly monitoring braking pressure. Initially marketed as an optional extra on top-of-the-range models, it is now a standard feature on almost all new cars.

TRAFFIC IN TOWNS
Consequences of the car being king

After the Second World War, the car rapidly became the main mode of transport in US cities, then in most other developed nations. The trend grew stronger in the 1960s, when car ownership became a central focus of most people's ambitions. Over time, traffic congestion became a problem that could not be ignored.

Gridlock in Paris
Cars jam the Place de la Concorde in the 1960s (right).

Busy junction
The interchange linking Paris's inner ring road and the A3 motorway at Porte de Bagnolet (below) handles around 300,000 vehicles a day.

In the immediate post-war years the car appealed to city-dwellers as a quick and convenient way of getting around. It was more comfortable than public transport and allowed whole families to travel together with ease. By the mid 1950s mass production had reduced manufacturing costs and purchase prices, making car ownership open to almost all.

From the 1960s on, the increasing reliance on cars began to affect the structure and look of cities, although this was more apparent in some countries than others. Efficient traffic management meant widening existing roads and constructing expensive infrastructure in the form of tunnels, expressways, ring roads and bypasses, as well as installing traffic lights, roundabouts, underpasses and restricted lanes. One of the prices paid was more limited pedestrian access. As more and more cars lined the pavements, parking facilities had to be provided, whether public or private, underground or above ground, and parking meters became a regular urban feature. The expansion of the highway system created new nuisances in the form of more noise, danger

Information overload
Signs cluttering a Tokyo lamppost provide a multiplicity of messages for motorists. Some road symbols have gone international, reflecting the increasingly cross-border nature of modern motor traffic.

and pollution, while the increased number of cars on the roads meant more hold-ups and bottlenecks. Measures to control things included traffic counts, the synchronisation of traffic lights and real-time flow management.

To lessen congestion, city authorities built out-of-town retail centres better suited to cars than the saturated city centres, with easy access from motorways. A downside was that these new shopping zones were rarely well served by public transport and over the years they have moved further and further out. To contain the urban sprawl around existing towns and cities, the authorities began to create new towns with roads designed to accommodate cars. None of these initiatives have done much to challenge the idea that everyone should have access to cars. Equally, it has become increasingly common for people to live in one town and drive to work in another.

A growing problem

In 1955 there were 67 million private cars around the world, a figure that almost trebled to 190 million by 1970, then more than

THE ALGEBRA OF TRAFFIC JAMS

In the late 1960s a German mathematician, Dietrich Braess, gave his name to a paradox affecting traffic in towns. He showed that adding an extension to a road network can actually increase journey times. As if in proof of his theory, in 1969 the city of Stuttgart sought to ease congestion in its centre by building a new road. In fact the hold-ups got worse. The problem was that drivers, considering their own needs and not the flow of the network as a whole, avoided the new road, fearing it would lengthen their journey.

Drive-in clinic
The phenomenon of drive-in facilities only really took off in the USA. Here, motorists and their families participate in a vaccination programme in 1960, queuing up to get their polio jabs without having to get out of their cars.

trebled again to 600 million by 2010. Experts are predicting 2.9 billion cars by the middle of the 21st century. The upward spiral suggests that the constant increase must eventually run up against limits. Far from freeing people, cars can end up imprisoning them in traffic jams and generating stress. The rise in car ownership may have introduced a new lifestyle since the 1950s, but it has also fostered unwelcome behaviour like dangerous driving and road rage. By the early 1970s nearly 8,000 people were dying as a result of accidents on Britain's roads each year, about 3,000 of them pedestrians. To reduce these numbers, new measures were introduced that included the use of breathalysers and radar speed checks.

Between 1973 and 1979, a succession of oil crises pushed up the price of petrol, forcing motorists to make economies. In response cars in Europe became smaller and less extravagant (American manufacturers were much slower to pick up the trend). In turn, the energy crisis made people newly aware of environmental problems in general, and in particular of the dangers of exhaust gases. As demands increased for something to be done, the automobile industry responded first, from 1974 on, by introducing catalytic converters, then four years later removed lead from petrol altogether.

Traffic-calming measures

By 1990, swayed by environmental concerns and the growing congestion of the streets, city dwellers and municipal authorities across western Europe were increasingly seeking alternatives to cars. By 1994 the Netherlands, Germany and other countries were introducing green zones where motor traffic was forbidden (except for emergency vehicles) and parking was increasingly kept to the edges of urban areas.

Cycle lanes were introduced to make cycling safer and encourage people to use bicycles, thereby helping to reduce car use. To an extent, success depended on the urban terrain. In the cities of Amsterdam, Copenhagen and Fribourg, up to 30 per cent of all journeys came to be made that way. In Britain Cambridge had great success in promoting cycling and many other centres – from London to Bristol, Brighton, Exeter and Lancaster – have taken steps to encourage people to get on their bikes.

Car-sharing is another initiative aimed at reducing the number of vehicles on the road. According to a 2011 survey, 20 per cent of the UK population were regularly sharing rides. To encourage the trend in the heavily polluted city of Athens, access to the city centre was restricted on alternate days to cars with number-plates ending in odd or even figures. Radical schemes for traffic reduction in the urban heart have been adopted in various European cities, focused on encouraging motorists to abandon their cars and use public transport to access the main shopping and cultural centres. Cities like Berlin and Rotterdam have closed off the central areas to vehicles considered heavily polluting.

The measure that best symbolises the desire of municipal authorities to cut city-centre traffic came into operation in Britain in 2003, when Mayor of London Ken Livingstone introduced congestion charging for vehicles entering a defined area in the centre of the capital. The result was a reduction of about 15 per cent in vehicles entering the charging

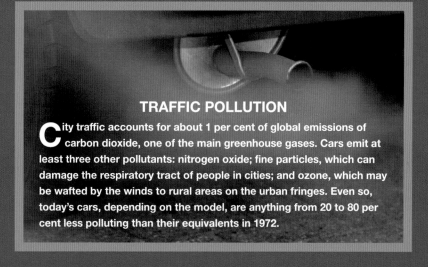

TRAFFIC POLLUTION

City traffic accounts for about 1 per cent of global emissions of carbon dioxide, one of the main greenhouse gases. Cars emit at least three other pollutants: nitrogen oxide; fine particles, which can damage the respiratory tract of people in cities; and ozone, which may be wafted by the winds to rural areas on the urban fringes. Even so, today's cars, depending on the model, are anything from 20 to 80 per cent less polluting than their equivalents in 1972.

Congestion cost
A sign at the start of London's congestion charging zone (below) reminds drivers going into the central zone of the need to pay the congestion charge – currently £10 a day between 7am and 6pm weekdays.

ALTERNATIVES TO THE CAR

Long neglected in favour of private car ownership, public transport has come back into fashion recently thanks to the energy crisis, rising petrol costs and growing awareness of environmental problems. Bicycles and motorcycles have both grown in popularity, but neither can take the place of urban mass-transit systems. Only public conveyances, which take up less space than cars and are less polluting, seem to offer a long-term solution.

Cleaner ways to travel
A single tramway, like this one (above) flanked by a cycle lane in Leipzig, can carry between 20,000 and 60,000 passengers a day, midway between the capacity of an underground line and a bus route

The big squeeze
The Lumeneo SMERA (below), an electrically powered car that came on the market in 2010, has two 20hp motors that provide 150km (90 miles) of motoring, but the batteries take five hours to recharge.

zone and a 10 per cent decrease in journey times as the remaining traffic moved more freely. This new form of taxation imposed financial penalties, rather than the threat of traffic jams, to dissuade motorists from driving in the heart of the city. The money collected is invested in transport infrastructure projects.

In the USA different cities have adopted different approaches depending on their own needs. New Yorkers are not much inclined to use their cars to get around the city centre, whereas residents of Los Angeles have little choice but to do so.

Less petrol, more electricity

The idea of removing vehicles altogether from city centres is unrealistic. A more credible alternative might be for motorists to either buy or hire electrically powered vehicles, although experts currently limit projections for this mode of transport to between 1 and 8 per cent of journeys. Rapid charging of batteries would be a major step forward. Perhaps the most likely prospect is the development of hybrid cars that address some of the problems posed by purely electric motors, notably the low power output and the need for frequent battery recharging. Fuel cells are regarded as a promising alternative, but will not be practical before about the year 2015. Meanwhile traditional engines are continually being improved, thanks to more effective particle filters, more fuel-efficient combustion and nitrogen oxide sensors used to reduce polluting emissions.

Progress needs to be made on two fronts if cities are to coexist comfortably with increasing levels of motor traffic. The vehicles themselves must become greener, and the people who drive them will have to modify their behaviour, suiting their mode of transport to the needs of each particular journey.

Daredevil sport
Dirt biking involves performing acrobatics on two wheels over earth or sand mounds.

Mountain bikes 1974

Designed for off-road use over rough terrain, mountain bikes have a number of distinguishing features including powerful brakes, suspension forks and a large number of gears, making it easier to climb slopes. They are made of supple materials that combine comfort with precision engineering.

In 1968 Geoff Apps, a motorbike trials rider from the Chilterns, set about adapting pedal-powered machines to meet the demands of his own chosen sport. But it was six years later that off-road biking really took off after a group of friends in California started hurtling along mountain tracks – a pastime that won the machines they rode the name of 'mountain bikes'. To meet the demands of the activity they devised a bike with a shortened, more compact frame provided with small wheels but large tyres, sometimes studded with knobs. Early models were rather cumbersome, but the designs rapidly improved.

Back to nature
Mountain biking has proved a popular attraction in Britain, drawing people to explore the wilder places of Wales, the Borders, Cumbria and the Highlands.

Going mainstream

In the early 1980s firms that manufactured touring bicycles also started making high-tech mountain bikes. Geoff Apps's company produced the first 29ers at this time, so-called for the 29-inch wheels. As the bikes became increasingly mainstream, readily available in high-street shops, a mountain biking community developed with their own specialist suppliers and magazines. A new extreme sport had been born, tempting urbanites out into the wilds of the countryside.

THE BMX CRAZE

Bicycle motocross, or BMX, first saw the light of day in the USA in the early 1970s. As the name suggests, the idea came from motocross racing, and the first BMX enthusiasts were children too young to have motorcycle licences who rode their bicycles over motocross trails. BMX racing takes place on courses laid out over rough terrain, typically between 300m and 500m long and strewn with banked corners and mounds that send the bikes flying through the air. BMX became a full Olympic sport for the first time at the Beijing Games in 2008.

INVENTIONS THAT CHANGED OUR LIVES

Rubik's Cube 1974

Originally called the Magic Cube, Rubik's Cube is a three-dimensional puzzle invented in 1974 by Erno Rubik, a Hungarian sculptor and professor of architecture. Each of the six faces of the cube is made up of nine squares, coloured white, blue, orange, green, red or yellow. A mechanism patented by Rubik in 1975 makes it possible to turn the separate lines of squares on different axes, thereby mixing up the colours. (The centre square on each face remains static, providing a stable core.) The aim then becomes to restore each face of the cube to a single colour. More than 350 million cubes have been sold around the globe, making the device the world's best-selling toy.

Competitive solvers
The first Rubik's Cube world championship was held in 1992. These competitors (below) were taking part in the 2007 event, which drew entrants from 32 countries to Erno Rubik's home city of Budapest.

CHAMPION SPEEDCUBERS

Born in Australia in 1995, Feliks Zemdegs set a new world speed record in early 2011 at the Melbourne Open, solving the puzzle in 6.65 seconds. Previous champions have included Dutchman Erik Akkersdijk, Japan's Yu Nakajima and Andrew Kang from the USA. Encouraged by the result, Akkersdijk took part in a televised broadcast in which he took just 90 seconds to rearrange the cube with his feet. Other enthusiasts have completed the challenge underwater in a single breath.

Catalytic converters 1974

Catalytic converters, which are now fitted to 85 per cent of all the world's cars, were first introduced in the USA by General Motors in 1974. They did not reach Europe until 1985, but were made obligatory by the European Union in 1993. The French chemist Michel Frenkel had in fact taken out a patent on a way of making exhaust gases less noxious as far back as 1909, but there was little interest in environmental measures at the time. It was only much later that car manufacturers developed a way of reducing the toxicity of exhaust emissions. The converter has a honeycome-like internal structure, inside a ceramic or steel support, creating small channels that enclose the catalysing metals – microparticles of palladium or platinum. By recombining atoms of nitrogen, carbon and oxygen, the metals reduce or eliminate toxic gases such as carbon monoxide, nitrogen oxides and uncombusted hydrocarbons, reducing their proportion to about 1 per cent of the whole and ejecting the balance in the form of carbon dioxide and water.

SMART CARDS – 1974
A computer in a strip of plastic

Invented by France's Roland Moreno, the smart card has become a part of all our lives. Bank cards, telephone cards, SIM cards, security passes, biometric passports – all use computer chips to provide and secure information.

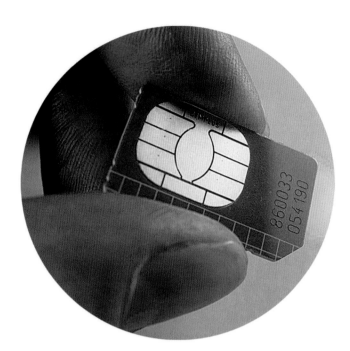

Taking his inspiration from a science-fiction hero who had a ring that gave him access to a spaceship, Roland Moreno had the idea of integrating a microchip into a signet ring. He called the project TMR, ironically referencing a Woody Allan film *Take the Money and Run*, for he imagined the ring as an Open Sesame banking device, portable yet provided with a memory that could be used to pay for purchases in shops. Moreno hoped, too, that it would combat fraud and hasten the advent of electronic money. He submitted his invention to the executives of various financial institutions, who encouraged him to develop it in bank-card form. The result, in 1974, was the smart card, which has since become an accepted part of daily life.

In the spirit of the times
The idea of integrating an electronic memory component in a credit card was very much in the air at the time. It had germinated in the minds of two German engineers, Jürgen Dethloff and Helmut Grotrupp, as early as 1967. The first patents were taken out by Kunitaka Arimura of Japan in 1970 and Paul Castrucci, a US citizen, in 1971, although neither went on to develop a working model. What changed in 1974 was that Moreno read a magazine article that informed him of the existence of PROM – Programmable Read-Only Memory – in the form of integrated circuits that the user could programme.

Nine years after the launch of memory cards came telephone cards, which blew a tiny fuse each time a unit was consumed. Created for France Telecom by Schlumberger, the 'pyjama card' (so called for its striped design) replaced coins as the operating medium for phone booths, which were spreading across France at the time.

Replacing magnetic strips
These early cards were passive, with no function other than to read and write information stored electronically in the chip's memory. That changed with the arrival of microprocessor cards, developed by Michel

Telephone card
A telephone card (above) issued by France's largest telecommunications provider, France Telecom. Made of PVC, such cards contain a chip that remembers the balance of units left in the account as well as the protocol needed to communicate with the card reader.

SMART CARDS – 1974

Ugon for Honeywell Bull in 1977. Roland Moreno described these as 'small, secure portable computers in a strip of plastic'. They were active, in the sense that they were able to do calculations and perform more than one task at a time. Smart cards operate like dormant computers, coming to life when they are inserted in a reader.

From that time on smart cards have steadily displaced magnetic-strip payment cards, which had been introduced in 1971. They have done so because they are faster to use, being able to effect transactions locally without the need to relay data back and forth to a central computer: the cards themselves contain information identifying the holder and the amount of credit available. Payees transmit this information to the bank and all the data transfer takes place between the card and card reader. In addition, users can identify themselves by entering a PIN (personal identity number) that the card itself checks against its own memory to determine whether or not to authorise the transaction. Security is thus reinforced, the more so as the microprocessor can also be used to encrypt data, making it inaccessible to anyone without the necessary cipher. The only way would-be fraudsters can beat the system then becomes to physically steal the card and the information needed to work it.

A SELF-TAUGHT INVENTOR

Born in Cairo in 1945, Roland Moreno first attracted notice as the inventor of a number of Heath Robinson-like devices designed for the sheer pleasure of their ingenuity. In 1968, while earning his living as a courier for a French news magazine, he kept the editorial team amused with a series of bizarre creations that included a version of heads and tails in which red and green lights started blinking when a button was pressed, and only one stayed on when the button was released. There was a machine that made matchsticks jump randomly on a stretched skin and a calculator designed to give the wrong result. His eccentricities inspired a sequence in a French film of 1970, Claude Sautet's *Les Choses de la Vie* (*The Things of Life*). In 1972 Moreno founded Innovatron, an organisation dedicated to 'selling ideas to people who don't have them'. After inventing the smart card, he took the precaution of taking out some 40 patents in 11 different countries. He has also written a number of books reflecting his quirky outlook on life, of which the best-known is *The Theory of Ambient Chaos*.

Inventor at work
Roland Moreno, inventor of the smart card, at work in his Paris home in 1988. He received France's highest award, the Legion of Honour, in 2009.

Five billion cards in use

From 1992 on smart cards equipped with microprocessors became the norm for banking purposes and their predominance has not yet been challenged. By 2008 more than 5 billion of the cards were in use around the world, 80 per cent of them equipped with microprocessors. Some 3.2 billion cards were in use in telecommunications and 650 million for banking.

Europe took the lead in smart-card development, but the concept quickly spread round the world, reaching South Africa, Russia and Japan in the 1990s, by which time Europeans were coming to regard the cards as the most common and

Swipe card
Contactless swipe cards, like this Navigo pass valid on rail and Metro services in and around Paris (left), are so called because they do not need to be fed into a card reader, simply passed near one. They use technology known as radio frequency identification (RFID), operating within an electromagnetic field generated in the immediate vicinity of the card reader.

DIFFERENT TYPES OF SMART CARD

There are three main types of smart card. Memory cards do not contain microchips and simply register information that can subsequently be modified – for instance telephone cards from which units are deducted each time the card is used. Cards that do contain microchips, such as bank cards with PIN numbers, can deny access to services the third time a wrong number is entered. Contactless smart cards, like London Transport's Oyster card, use electromagnetic fields and radio-frequency identification technology to convey information by swiping the card across a reader; they are popular for providing access to public transport because they do not need to be inserted into a card-reader.

SMART CARDS – 1974

THE NEED FOR DATA SECURITY

Card issuers have taken time to learn the importance of protecting data contained on smart cards from improper access. In 1998 the French social services introduced the Carte Vitale to replace the country's existing social-security ID card. It was used for healthcare payments, but it was soon proved insufficiently secure because users did not have to provide authenticating proof of identity to access the information that the chip contained. In addition, the data was not encrypted, making it relatively easy for anyone with the know-how to copy the cards and change the information they contained, including the amount of money to be reimbursed. A replacement card addressing these problems was introduced gradually from 2006.

Electronic payment
Today, there are various electronic ways of paying for goods without using cash, including bank cards, stored-value cards for small transactions and, in the near future, mobile payment, which is already available in Japan and is being tested in some European cities. This involves using mobile phones containing integrated circuits as payment cards; as soon as the phone is 'recognised' by the storekeeper's card reader, the transaction can begin.

convenient way of paying for consumer goods and services. Only the USA held out against the trend, preferring to stick with magnetic-stripe cards despite their lower security.

Smart cards are now ubiquitous, and they have many different uses. Chip'n'pin cards can be used to make purchases in stores and collate data about people's purchasing habits. SIM cards – the initials stand for Subscriber Identity Module – that are inserted into mobile phones let people communicate with one another, identifying the subscriber and containing data that includes the telephone number and all relevant information about the operating company and individual subscription needed to make a connection to the mobile network.

With the advent of biometric passports, smart cards are being used to identify people. Their many functions include providing access to transportation systems and serving as security passes, letting people into business premises or casinos. Medical services in some countries make use of them (see box, above) and they play a part in the leisure industry (where they are used to decode pay TV channels) and increasingly serve as keys for some makes of car. Even if smart cards still cannot make doors appear as if by magic, like the original 'Open, Sesame!', Roland Moreno's invention has many amazing functions waiting to be explored.

MINIATURISATION
Welcome to the nanoworld

The story of miniaturisation began with the invention of the transistor in 1947, then accelerated following the arrival of integrated circuits and microprocessors. Extending over several decades, it launched the modern electronic era.

At the end of the Second World War, the goal of miniaturisation could be summed up as getting more from less: more components from less raw materials, more performance with less inconvenience, more power from less energy. And more profit from reduced investment, for the electronics revolution was driven above all by economic considerations. Radios, telephones, TV sets and computers all started off big and cumbersome, then shrank until, in time, they could be held in the palm of the hand.

Ever-more-powerful chips

These changes would never have happened without the development of integrated circuits by Jack Kilby at Texas Instruments in the late 1950s. By manually linking together a number of transistors, he created the first chip. It was this that set off the miniaturisation revolution. Integration now became a key concern of the nascent high-tech sector, expressed in terms of the number of transistors that could be incorporated on a given area of silicon. The fastest progress was made in computing, where the multiplication of interconnections on one circuit paved the way for the development of increasingly complex systems. Over the course of the 1970s the number of components that could be compressed onto a single chip increased by a factor of 12.

Technological innovations

Since the 1960s the products of miniaturisation have become very much a part of everyday life. Transistor radios – small, easily portable and powered by batteries – revolutionised listening habits. TV sets underwent a similar process: needing 100 times fewer transistors than a basic calculator, they had been shrunk to the size of shoeboxes by the mid 1970s – at least for people willing to put up with a smaller screen. Music became easy to carry around thanks to the development of reduced-size audio-cassette players. In 1966 tape recorders were substantial machines that needed stands the size of bedside tables, but it was not long before they had been reduced to a fifth of their former dimensions. The first commercial pocket calculator appeared in 1972.

Meanwhile industry was struggling to get a grip on the economic levers that could deliver products for a mass market. Leaving aside tensions raised by the Cold War and the space

MINIATURISATION

MASTERS OF THE ART

Not only are the Japanese champions of miniaturisation, they are also fervent adepts of reductionism, a philosophy whose origins are lost in the early years of the nation's history. Bonsai trees exemplify the uniqueness of Japanese taste in this respect when compared with the rest of the world. The technological drive of the post-war years provided an opportunity that Japan did not let slip. Manufacturers quickly came to excel in reducing the size of electronic components and developing ever-smaller devices. Competing directly with the USA, Japan managed to establish itself as the market leader in consumer electronics. Sony were putting transistors in radios in 1952 and launched one of the earliest portable radios three years later. In 1979 Western competitors were taken by surprise by the appearance of the Sony Walkman, even though a German, Andreas Pavel, had produced a personal stereo called the Stereobelt seven years earlier. Following the launch of the first personal computers in 1981, Japanese firms played a central role in developing laptops.

Embracing downsizing
A woman tests one of the first miniature radio sets in the 1950s (left). By the mid 1970s this small TV set (above) was on the market: it ran off the mains but could also be operated with rechargeable batteries. The recommended viewing distance for optimal picture quality was about half a metre.

Mini-camera
It is now possible to take photographs with devices integrated into USB flash drives.

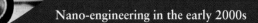

Nano-engineering in the early 2000s
The world's smallest fork-lift truck (left) was built by a German engineer at the Mainz Institute of Microtechnology.

MINIATURISATION

MICROCOPTERS

The world of modelling embraces many passionate enthusiasts filled with wonder at the performance of their toys for grown-ups. It is easy to understand why, considering the degree of technological wizardry that goes, for example, into getting miniature craft to fly. Just 17cm (6.5in) long and weighing 10g, the Picco Z microcopter wheels about with a surprising degree of stability even in very confined spaces. The Black Hornet nanocopter has all the attributes of a full-size helicopter, but weighs only 15g. Powered by an electric motor, it has four rotors, carries a digital camera, and flies at speeds of up to 30km/h (19mph). These pilotless craft are designed for reconnaissance on military and civilian observation missions.

Tiny flying machines
The lightest helicopter in the world (left), weighing just 1g, went on display at the 7th European Micro Air Vehicle Conference in September 2007.

Less is more
The model below holds in front of him a 6GB hard disk manufacturered in 1999, an increase in capacity by a factor of 1,500 compared to the 4MB 1984 IBM disk held at his back.

race, the USA, Europe and Japan also confronted the Eastern Bloc nations in the matter of miniaturisation.

Enter the microprocessor

Intel put the first microprocessor, the 4004, on the market in 1971 and close on its heels brought out one of the earliest home computers. It was a decisive turning-point for the IT industry, which over a period of 25 years had seen computers shrink from the size of ENIAC – a 30-tonne machine covering 63m^2 of floorspace – to a device that could fit on a desktop. The era of microcomputing had finally got under way.

The drive to miniaturise soon affected other sectors. In the last years of the 20th century, microprocessors were employed to make a wide range of devices less cumbersome, from TV sets to hi-fi equipment. The first cellular phones made their appearance in the late 1970s, around the size of a housebrick and weighing almost 1kg. They would soon be on their way to becoming the compact devices we use today. Sony introduced the first camcorder in 1982.

MINIATURISATION

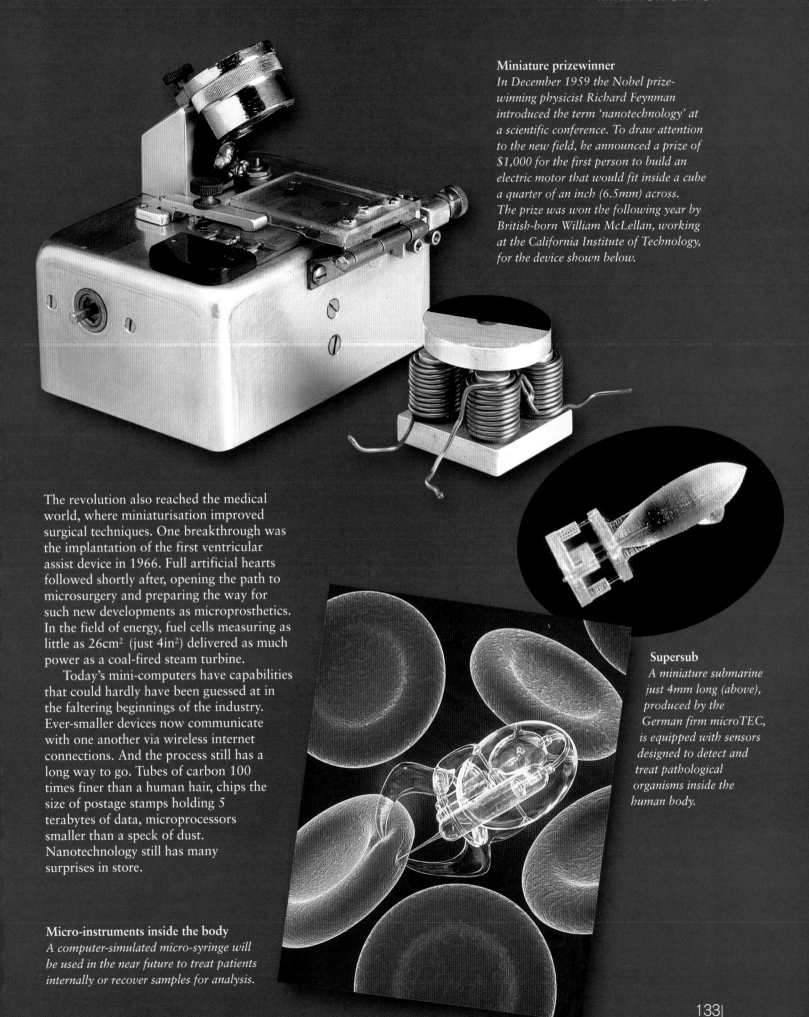

Miniature prizewinner
In December 1959 the Nobel prize-winning physicist Richard Feynman introduced the term 'nanotechnology' at a scientific conference. To draw attention to the new field, he announced a prize of $1,000 for the first person to build an electric motor that would fit inside a cube a quarter of an inch (6.5mm) across. The prize was won the following year by British-born William McLellan, working at the California Institute of Technology, for the device shown below.

The revolution also reached the medical world, where miniaturisation improved surgical techniques. One breakthrough was the implantation of the first ventricular assist device in 1966. Full artificial hearts followed shortly after, opening the path to microsurgery and preparing the way for such new developments as microprosthetics. In the field of energy, fuel cells measuring as little as 26cm² (just 4in²) delivered as much power as a coal-fired steam turbine.

Today's mini-computers have capabilities that could hardly have been guessed at in the faltering beginnings of the industry. Ever-smaller devices now communicate with one another via wireless internet connections. And the process still has a long way to go. Tubes of carbon 100 times finer than a human hair, chips the size of postage stamps holding 5 terabytes of data, microprocessors smaller than a speck of dust. Nanotechnology still has many surprises in store.

Supersub
A miniature submarine just 4mm long (above), produced by the German firm microTEC, is equipped with sensors designed to detect and treat pathological organisms inside the human body.

Micro-instruments inside the body
A computer-simulated micro-syringe will be used in the near future to treat patients internally or recover samples for analysis.

THE DISCOVERY OF LUCY – 1974
An Australopithecine opens a window onto our origins

The world's most celebrated hominid fossil – named Lucy by the archaeological team who found her – was dug up in the Afar region of Ethiopia on 24 November, 1974. Since that time Lucy has lost her claim to be humankind's oldest known ancestor, but even so still has a special place in our genealogical tree.

Toumaï's skull
Known to be about 7 million years old, Toumaï's cranium (below) has the brain size of a chimpanzee but exhibits characteristics that link it to human evolution (below).

'We've found it! An almost complete skeleton!', exclaimed Tom Gray, an American student announcing the discovery of Lucy to colleagues working a few metres away. The place was the Afar Depression in northwest Ethiopia where Gray was working on an archaeological dig under the direction of Donald Johanson from the USA and Yves Coppens and Maurice Taïeb of France. As they hastened to the spot, they were dumbfounded by what they saw: lying out in the open were ribs and vertebrae from the skeletal remains of what looked like a petite woman. They were in an excellent state of preservation, indicating that the body had been rapidly buried. Taken together with the nature of the surrounding sediments, this suggested that she had probably been swept to her death by a current of water. Back in camp that evening the team nicknamed her Lucy after the Beatles' song *Lucy in the Sky with Diamonds*, which was being played incessantly on a cassette player at the time.

Walking on two legs
Ever since her discovery, Lucy's remains have been considered of prime importance for our understanding of the origins and evolution of the human race. First, at the time it was the most complete hominid skeleton yet discovered; among the 52 bone fragments were parts of the skull, a mandible, many bones from the limbs and thorax, a femur and part of the pelvis. Analysing the latter allowed palaeontologists to confirm that Lucy was indeed female, and by measuring the skeleton they established that she stood just a shade over a metre tall (3ft 5in). Her diet, deduced from the state of her teeth, consisted mainly of fruits and leaves from trees, plus bulbs and roots dug up from the soil. Most significantly of all, the curve of her spine, together with the shape of the pelvis and femur and the positioning of the head, indicated that Lucy was fully bipedal, walking on two legs rather than all fours. Even if her gait had been somewhat lumbering and primitive, this still linked her to the human line of descent.

A new species
On the basis of her anatomy, which differed significantly from that of other human ancestors known at the time, Lucy's discoverers decided in 1978 to assign her to her own particular species, *afarensis*, within the genus *Australopithecus*. This last term, coined in 1924, designated a family of bipedal hominids with archaic anatomical features who disappeared from the Earth about 2 million years ago. In Lucy's case, geologists were able

> **WHO IS THE OLDEST HUMAN ANCESTOR?**
>
> Three fossils are currently in competition for the hotly contested title of the oldest-known ancestor of the human race – the hominid remains that are closest to the start of the evolutionary line that led to humankind. There is no question that the oldest remains yet discovered belong to *Sahelanthropus tchadensis*, nicknamed Toumaï, which were found in Chad in 2001 – a recent dating has confirmed that the skull fragments are around 7 million years old. But what is disputed by some is whether the fragments belonged to an early predecessor of the great apes.
> Next comes *Orrorin tugenensis* at 6 million years old. First uncovered in Kenya in the year 2000, this species walked on two legs like humans, as the presence of long, straight femurs attests. Finally *Ardipithecus ramidus*, which lived in Ethiopia 4.5 million years ago, had teeth exhibiting mixed anatomical characteristics, midway between a monkey and a man. There is little agreement among experts as to the position that each of these three specimens should occupy on the genealogical family tree.

Home to hominids
Stretching some 6,000 miles from the Red Sea to the Zambezi River, the dry Great Rift Valley provided a favourable environment for the fossilisation of hominid remains, many of which have been found there. Lake Natron (above) is in a section of the valley in northern Tanzania.

to date the volcanic tuff rocks beneath the skeleton to 3.2 million years, making Lucy the oldest known representative of the human evolutionary line found to that date and winning her the nickname of 'grandmother of the human race'. The date was obtained by the potassium–argon method, which allows scientists to determine the age of a sample by measuring the relative concentrations of these two isotopes, given that potassium decays to form argon over long periods of time.

Dethroning Lucy

Since the time of her discovery, Lucy and other *Australopithecus afarensis* specimens unearthed after her have been joined by other early hominid remains disputing their claim to be the oldest human ancestors. A dozen different species are now recognised as predating the appearance of the *Homo* genus some 2 million years ago – in other words, beings exhibiting clearly human anatomical features. Some of these potential ancestors, discovered in the 21st century, are far older than Lucy (see box, left).

Even so, her skeleton remains special for palaeontologists. Its excellent state of preservation provides a point of comparison with the more rudimentary remains of freshly discovered fossils, which continue to emerge year on year. As evidence of her importance, Lucy's bones are preserved in a safe in the Ethiopian capital Addis Ababa, protected from the public gaze, compelling researchers to do most of their work from plaster casts.

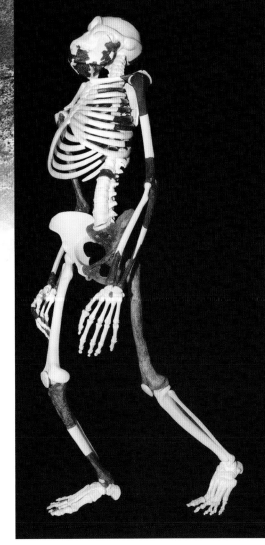

Reconstituting Lucy
Working from the evidence of the bones preserved at the National Museum of Ethiopia – shown in brown on the skeleton (above left) – the Finnish taxidermist Eirik Granqvist produced the life-size model of how Lucy might have looked (above right).

COMPETING THEORIES

In the early 1980s the anthropologist Yves Coppens put forward a theory that sought to explain how the human race evolved from the most ancient fossil forms. Nicknamed East Side Story, it postulated that our ancestors gradually adopted a vertical posture as an adaptation to an increasingly deforested, savannah-like environment to the east of the Great Rift Valley, an impassable geological barrier that started to take shape 10 million years ago. According to the theory, the ancestors of today's great apes evolved in the forests that survived to the west of the rift. The discovery of the fossil named Toumaï in Chad, west of the rift, cast doubt on the theory. Currently researchers incline to a more complex account of our origins, by which several species of primates exhibiting different forms of bipedalism coexisted in surroundings that were half forest and half savannah. One of these evolved towards the genus *Homo* without the clean break from the other species that had previously been suggested.

ENDORPHINS – 1975
The body's own opium

Passing the pain barrier
Ethiopia's Haile Gebrselassie is one of the all-time giants of distance running, the setter of 27 world records. He was Olympic champion over 10,000m in both 1996 and 2000 and subsequently turned to the marathon. Here he is winning the 10,000m event at the 1995 World Championships held at Gothenburg in Sweden. He completed the race in 27 minutes and 12.95 seconds.

As endurance athletes know, a sensation of euphoria kicks in after a prolonged period of effort, wiping out the pain. This 'runners' high' is produced by endorphins – hormones related to morphine that are released in the brain. For the scientists who discovered the existence of endorphins, it was no easy task to get others to acknowledge that the body secretes its own version of opium.

In the early 1970s, when the analgesic benefits of morphine had already been recognised for more than 150 years, several groups of researchers made a disturbing discovery: sensory receptors specifically coded to respond to morphine and to its derivatives, the opiates, are present in nerve cells. It is now known that they are abundant, particularly in the limbic system, the area of the brain associated with sensations of pleasure.

Touring the abattoirs

At first scientists were hard pressed to explain the existence of receptors coded for a plant-derived substance like morphine in animal cells such as neurons. Two Scottish researchers, Hans Kosterlitz and John Hughes, developed a daring hypothesis: what if the body produced its own morphine? In face of widespread scepticism, the two scientists decided to test their theory. Hughes toured the abattoirs of Aberdeen in search of pigs' brains. From this raw material he and Kosterlitz were able to identify two tiny peptides, each composed of five amino acids, that mimicked the analgesic effects of morphine and were linked to the opioid receptors. They named them enkephalins (from the Greek for 'in the head'), publishing their findings in *Nature* in 1975.

Close on their heels, other teams, including those headed by Andrew Schally and Roger Guillemin, discovered other, longer peptides with similar properties. Guillemin a French-born US citizen, had already established a reputation for his work in neuroendocrinology, a new discipline studying the interactions between hormones and the nervous system. He and Schally shared the Nobel prize for medicine in 1977.

ENDORPHINS – 1975

Natural high
To secrete endorphins, the body must maintain physical effort, such as swimming, for at least 30 minutes, always maintaining a steady rhythm. Production increases as long as the exercise continues and can be multiplied by a factor of five 30 to 45 minutes after the activity stops. The circular image (left) shows polarised crystals of beta-endorphin as seen under a microscope.

> **MULTIPLE EFFECTS ON THE ORGANISM**
>
> Besides their euphoric and analgesic qualities, endorphins have other properties that contribute to well-being, including reducing stress, notably by controlling respiration, stimulating sleep and combating depression. They are also implicated in the passage of food through the intestines and in the adjustments of the immune system. Over-production of endorphins can cause headaches, nausea and vomiting. Cases of endorphin dependence have been diagnosed in some long-distance runners.

Understanding pain

The discovery of endorphins was revolutionary in the light it cast on the inner mechanisms of pain as well as on our understanding of human drug dependence. The presence of these neurotransmitters in greater or lesser quantities went some way to explaining different people's sensitivity to pain, and also the varying response of individuals over the course of time, influenced not just by the nature of the discomfort itself but also by their own psychological and physical state.

Pain makes itself felt when the natural capacity of endorphins to block the pain receptors is overwhelmed; it disappears when the opiate arrives to saturate the unblocked receptors. This system also helps to explain withdrawal symptoms: abrupt removal of a morphine product leaves the opioid receptors unoccupied, causing an overpowering craving for its restoration. Twenty or so endorphins have now been identified, classified in three different groups according to their affinity to three types of receptors.

SILICON VALLEY
The technological heart of the planet

In 1971 Don Hoefler, a journalist at *Electronic News*, published a series of articles on Intel's launch of the first microprocessor. The title he chose for the pieces was 'Silicon Valley, USA', silicon being one of the principle materials used to make electronic components. From that time on the term became synonymous with the technology hub to the south of San Francisco.

Silicon Valley's success story started with an initiative taken in 1951 by the trustees of Stanford University, located in Palo Alto, near San Francisco. Facing financial difficulties, they decided to rent out part of the large campus surrounding the venerable institution on 99-year leases to high-technology businesses. This masterstroke created the Stanford Industrial Park, which quickly attracted General Electric and Eastman Kodak, among other clients.

In 1957 eight young executives with degrees in electronics quit the Shockley Semiconductor Laboratory (SSL), founded close to the university by William Shockley, co-inventor of the transistor. The 'eight traitors', as Shockley came to call them, jointly formed their own business, Fairchild Semiconductors, which marketed the first silicon-based transistors.

Extraordinary growth

In 1961 a New York banker named Arthur Rock was so taken with the dynamism of this corner of California, which had previously been known mainly for fruit-growing, that he based one of the first venture-capital funds there. Together with cutting-edge research from a top university and a spirit of risk, this provided all the conditions needed to create a nexus of brain power. Silicon Valley became a paradise for start-ups – firms that anticipated such significant trends as the defence-industry build-up in the 1960s, integrated circuits in

Small beginnings
Above: An informal meeting at Fairchild Semiconductors in the early 1960s.
Left: The modest garage in Palo Alto where, in 1939, William Hewlett and David Packard, engineering graduates from Stanford University, set up their own company. The firm, still bearing their joint names, is now one of the world's most important electronics manufacturers.

PARC – HOME OF INNOVATION

Created by Xerox in 1970, PARC (the Palo Alto Research Center) quickly became one of the valley's cradles of innovation. Thanks to the quality of the scientists employed there, it gave birth to the computer mouse, the Ethernet protocol, the laser printer and graphic computer interfaces employing icons, among other innovations. Steve Jobs, the founder of Apple, was inspired by PARC's work to conceive the first Macintosh home computer in 1979.

THE SILICON VALLEYS OF EUROPE

Many places in Europe have sought to emulate Silicon Valley's formula for success, from Oulu in Finland, some 400 miles north of Helsinki, to Tres Cantos in the suburbs of Madrid. Scotland has its own 'Silicon Glen' in the region between Edinburgh, Dundee and Glasgow – the name mimicked 'Silicon Fen', a title already applied to a cluster of high-tech start-ups around Cambridge in eastern England. In 2010 Prime Minister David Cameron publicly backed the development of the East London Tech City, located between Old Street and Stratford, home to the 2012 Olympic Park.

the 1970s, personal computing in the 1980s and the Internet in the century's last decade.

Of 45 US firms producing semi-conductors set up between 1959 and 1976, only five were outside Silicon Valley. In time the place became a nerve centre of the global high-tech economy, attracting a growing number of the best minds from around the world to concerns like Xerox PARC (see box, left), Hewlett-Packard, Intel, Apple, Cisco, Microsoft and Google.

With the coming of the Internet, the growth of Silicon Valley became even faster, raising the state of California to sixth place in the league table of world economies; Silicon Valley alone would have ranked 20th, had it been a nation. But then came the bursting of the dot.com bubble in 2001, which hit the Valley hard. Start-up financing temporarily dried up.

Even major concerns like Cisco and Hewlett Packard, carried away by the wave of double-digit growth and crazy financing, staggered under the blow.

Today Silicon Valley is home to 30,000 different businesses employing more than a million high-tech workers. It is currently betting on green technology and initiatives such as electric cars, solar panels and wind pumps to counteract the after-effects of the 2008 financial crisis and face down competition from other US centres located in Texas, Colorado and New York State – not to mention the challenge from low-cost producer nations like China and India. Investment in the sector doubled in 2008, and the number of jobs created by the new industries jumped by 23 per cent.

High-tech landscape
Silicon Valley spreads out around the shores of San Francisco Bay (top). Its main city, San Jose, is surrounded by a number of smaller communities, among them Redwood City, the headquarters of Oracle (above left), a leader in business software.

SNAPSHOT OF SAN FRANCISCO
A non-conformist 1960s city

In the 1960s San Francisco had the enduring charm of a metropolis built to a human scale, a *feng shui* city living in harmony with its site. It was at the heart of the protest movements that flourished at the time.

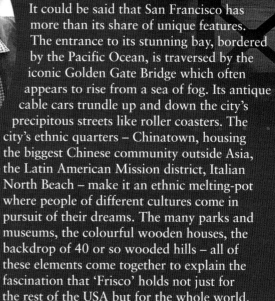

Sounds of protest
Joan Baez was almost as well known for her protest activities as for her singing voice. On a visit to Berkeley University in San Francisco in 1963 she made a public demand for free speech and the abolition of censorship. The photograph shows her performing with Bob Dylan to 200,000 protesters who converged on Washington DC in 1963 in support of civil rights for African-Americans.

It could be said that San Francisco has more than its share of unique features. The entrance to its stunning bay, bordered by the Pacific Ocean, is traversed by the iconic Golden Gate Bridge which often appears to rise from a sea of fog. Its antique cable cars trundle up and down the city's precipitous streets like roller coasters. The city's ethnic quarters – Chinatown, housing the biggest Chinese community outside Asia, the Latin American Mission district, Italian North Beach – make it an ethnic melting-pot where people of different cultures come in pursuit of their dreams. The many parks and museums, the colourful wooden houses, the backdrop of 40 or so wooded hills – all of these elements come together to explain the fascination that 'Frisco' holds not just for the rest of the USA but for the whole world.

The psychedelic revolution
The city, which first rose to prominence in the California Gold Rush of 1849, is undoubtedly the most European of the conurbations on the West Coast of the USA. When much of the nation remained puritan in outlook, extreme lifestyles cohabited there in a climate of tolerance. In the early Sixties it became a magnet and a refuge for people seeking the freedom to express themselves. Timothy Leary,

Symbol of the city
San Francisco's cable cars first went into service in 1873. Today three lines are still running, primarily serving tourists.

Music heaven
Rock promoter Bill Graham opened the Fillmore West venue in 1968. At the time he was still running the original Fillmore Auditorium in San Francisco and the Fillmore East in New York. The poster (above) advertised a 1969 concert featuring Santana and the Youngbloods.

Watery cityscape
The Golden Gate Bridge is not the only impressive bridge in San Francisco. The Bay Bridge (above left) was built in 1936 to join San Francisco with Oakland, the largest and most expensive bridge of its time. By 1958 the lower deck, originally for trains, had been converted for vehicles. Today the bridge carries ten lanes of traffic – five on each deck.

WAITING FOR THE BIG ONE

San Francisco's residents know that the San Andreas Fault, which runs parallel with the US West Coast for more than 600 miles, could open up at any time, triggering a cataclysmic earthquake like the one that struck in 1906. Measuring 7.8 on the Richter scale, the worst quake in California's history shook the city for 48 terrifying seconds, killing almost 3,000 people and leaving 250,000 more homeless. Scientists reckon there is a 62 per cent probability that the Big One will strike the San Francisco Bay area again before 2032.

a professor sacked from Harvard for his advocacy of the drug LSD, proclaimed the advent of the psychedelic revolution there in 1964, with backing from Allen Ginsberg, Jack Kerouac and William Burroughs, the literary figureheads of the Beat Generation. Psychedelia in turn engendered the hippie movement, spreading first across the USA and then across the Atlantic to Europe.

In the 'Summer of Love' of 1967, tens of thousands of young people opposed to US foreign policy were nonetheless liable to find themselves drafted to fight in Vietnam. Many converged on the city from all parts of the USA. The party got under way on 21 June

141

with a celebration of the summer solstice, then continued in a carnival atmosphere to the sound of rock bands and Tibetan chants. The Diggers, who took their name from a group of 17th-century English radicals who had sought to live without money or private property, distributed free food, turning the event into a gigantic picnic.

Hippy Haight-Ashbury

In the mid 1960s thousands of hippies descended on a working-class neighbourhood between the Buena Vista and Golden Gate parks around the intersection of Haight and Ashbury streets, attracted by the cheap rents. What they wanted to create was a communal lifestyle. A hit song of the day encouraged visitors to the city to wear flowers in their hair – many took it literally. The prevailing ideology was to be close to nature, so hair grew long and men rarely shaved. The district resounded to chants of Hare Krishna and iconic local bands like the Grateful Dead, Jefferson Airplane, and Big Brother and the Holding Company, featuring Janis Joplin.

In a joyous Utopian bazaar somewhere on the margins of the system, all sought to liberate themselves from bourgeois convention, rejecting capitalism for its dehumanising materialism and preaching free love. Soon, though, delinquents and disturbed individuals of every kind sought refuge in 'Hashbury', as it became known, turning it into a drug-users' ghetto where violence was rife. Between 1967 and 1969, LSD and marijuana gave way to heroin as the drug of preference, and a number of apprentice hippies metamorphosed into addicts and tramps. By 1970 the centre of the counterculture – today a beacon of yuppified bourgeois bohemianism – had been transformed into an urban jungle controlled by dealers. In San Francisco at least, the hippie movement was running out of steam.

HERALD OF THE COUNTER-CULTURE

Allen Ginsberg (1926-97) was one of the most influential figures of the Beat Generation. Brought up on the East Coast, he arrived in San Francisco in 1953, creating a sensation two years later with a reading of his long poem *Howl*. The work was an impassioned denunciation of an America given over to complacent materialism, championing instead political revolt, freedom, sex and drugs – all of which would be embraced by US youth in the 1960s. Ginsberg has been credited with coining the term 'Flower Power'.

In-tune trio *Allen Ginsberg (on the left) in Golden Gate Park in January 1967, performing with Maretta Greer and Gary Snyder at the giant happening known as the Human Be-In.*

Gay leader *Harvey Milk on 26 June, 1978, taking part in the seventh annual gay freedom parade staged in San Francisco.*

The Castro – the gay quarter

Close by Haight-Ashbury is the Castro, famous as a gay neighbourhood. San Francisco has had a self-aware homosexual community since the Second World War, when civilians and military personnel in transit found the city offered an environment more friendly to their way of life than other parts of the nation. In 1977 Harvey Milk, a declared homosexual, was elected to the city council as a Supervisor, becoming the first openly gay man to hold a major elected office in the USA. On 27 November, 1978, he was assassinated minutes after George Moscone, the city's mayor, had also been shot dead. The subsequent sentencing of Dan White, the confessed killer of both men, to an unexpectedly light jail term for manslaughter provoked rioting in the city. San Francisco's gay community would be under the spotlight again when AIDS struck in the 1980s.

Businessmen and windsurfers

In 1951 Stanford University created the prototype industrial park attracting high-tech companies. Today the campus continues to

SNAPSHOT OF SAN FRANCISCO

AN INTERACTIVE MUSEUM

The Exploratorium opened its doors in San Francisco in 1969. Dedicated to the sciences, it was the brainchild of the physicist Frank Oppenheimer, a brother of Robert, the father of the atomic bomb. The museum aims to show science as it is being created rather than merely staging chronological exhibitions of past discoveries. Visitors are encouraged to take a hands-on approach, manipulating the exhibits to get the intended effect. Many other similar institutions have since adopted the same principle, among them London's Science Museum, the Museum of Science and Industry in Manchester, the Glasgow Science Centre and Techniquest in Cardiff.

attract the best students, many of them hoping to forge a career in Silicon Valley, a suburban development zone whose prodigious growth has swept away the orchards that once filled Santa Clara County to the south of the city. Among the many state-of-the-art enterprises there is NASA's Ames Research Center, a leading player in the US space programme.

The bright young minds who work in Silicon Valley's cutting-edge firms occasionally escape from their computers and their dreams of making millions to ride the waves on surfboards and windsurfers. Others try to re-create the pleasures of surfing on land by taking up skateboarding.

San Francisco financial district is dominated by a Modernist architectural marvel, the Transamerica Pyramid, completed in 1972. The summit reaches 260m (850ft) skywards and the building is designed to withstand seismic shocks. Its foundations extend 15m (50ft) underground and the structure above this can sway with the movement of the earth in order to avoid collapse.

San Francisco still retains an atmosphere that marks it off from the rest of the USA. Its counter-culture leanings are reflected in its politics: the city has become a gathering-place for pacifists of all stripes. And its inhabitants continue to cultivate an art of living unlike any other in the world.

Sky-scraper
The Transamerica Pyramid points skyward behind a flat-iron apartment block (left). Designed by the architect William Pereira, the Pyramid has become another instantly recognisable symbol of the city.

Pacific rim
A surfer rides a giant wave off the coast near San Francisco. Ocean Beach runs for 3 miles along San Francisco's shoreline, providing fantastic opportunities for serious surfers.

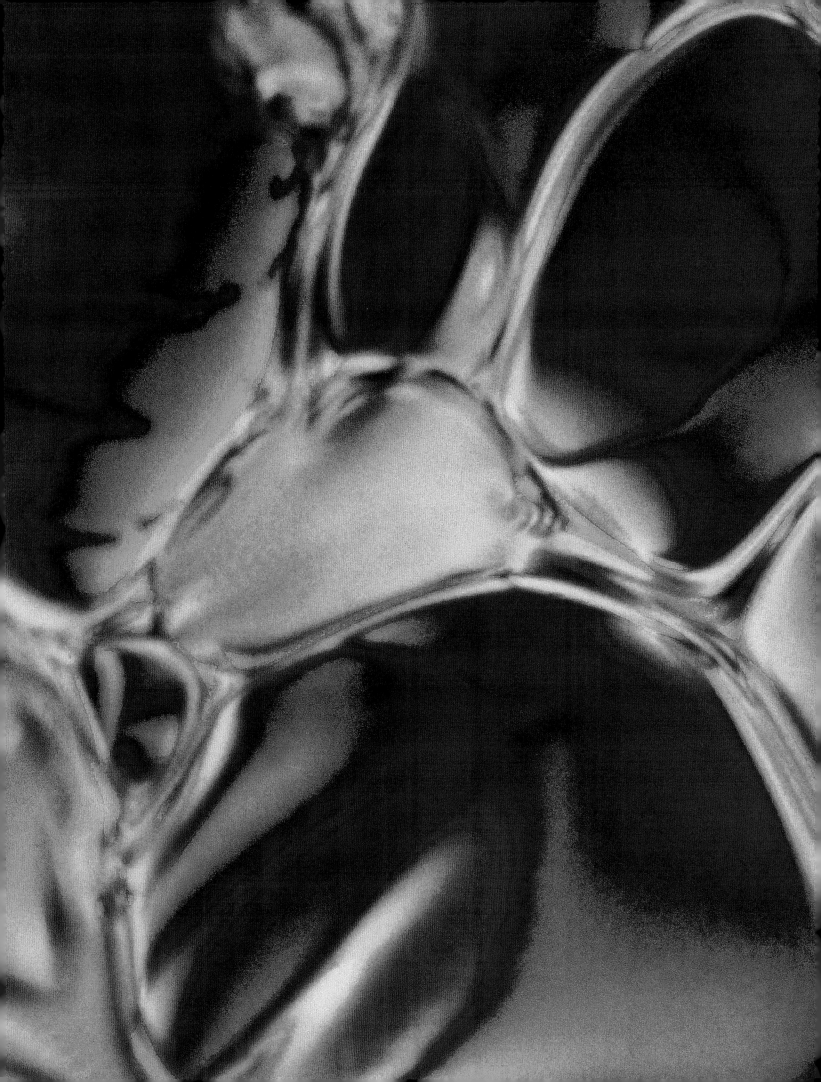

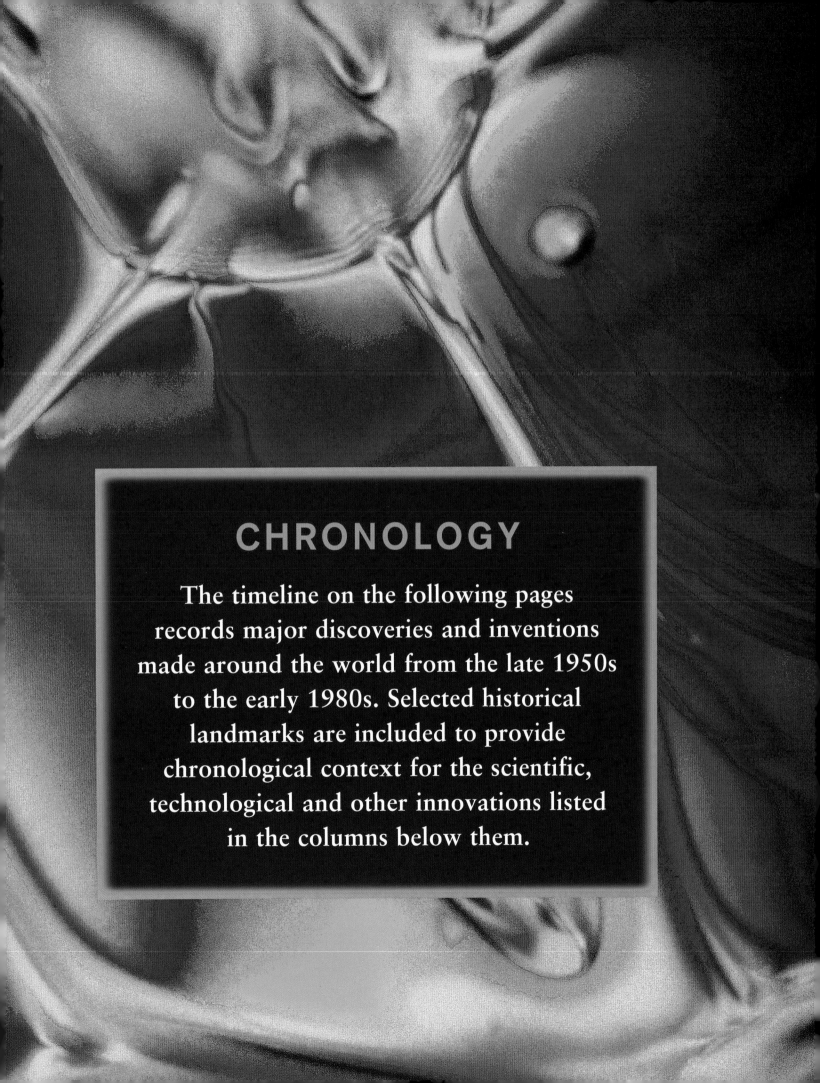

CHRONOLOGY

The timeline on the following pages records major discoveries and inventions made around the world from the late 1950s to the early 1980s. Selected historical landmarks are included to provide chronological context for the scientific, technological and other innovations listed in the columns below them.

CHRONOLOGY

1955

EVENTS
- The Warsaw Pact is signed, a treaty of mutual assistance linking the Soviet Union with Albania, Bulgaria, East Germany, Hungary, Poland, Romania and Czechoslovakia (1955)
- The Bandung Conference brings African and Asian states together to condemn colonialism (1955)

INVENTIONS
- Narinder Kapany develops optical fibre
- Christopher Cockerel takes out a patent on the hovercraft
- Synthetic diamonds are produced on an industrial scale
- Louis Essen builds the first accurate atomic clock, employing the caesium standard
- Ole Kirk Christiansen of Denmark manufactures the first Lego bricks
- Switzerland's George de Mestral patents the Velcro hook-and-loop fastening system

1956

EVENTS
- De-Stalinisation gets under way in the Soviet Union, opening the path to a policy of coexistence (1956)
- Russian troops put down an anti-Soviet uprising in Hungary (1956)
- Tunisia and Morocco gain independence (1956)
- President Nasser of Egypt nationalises the Suez Canal (1956)

INVENTIONS
- Victor Mills of the USA produces Pampers disposable nappies
- Art Ingels builds the original go-kart, using a lawnmower engine
- The oral contraceptive pill, developed by Gregory G Pincus, undergoes successful clinical trials
- The introduction of computer-assisted machine tools marks the start of a new era of automation that will revolutionise working practices in factories around the world
- Willem Kolff, a Dutch doctor working in the USA, develops artificial kidneys while working on renal dialysis
- The US toy manufacturer Mattel develops the Barbie doll®

▼ Assembling Sputnik 1 in 1957

▼ Radio-cassette player

◄ Tetra Pak carton

CHRONOLOGY

1957

- The Treaty of Rome establishes the EEC (1957)
- Mao Zedong launches China's Great Leap Forward (1958)
- Charles de Gaulle becomes president of France (1959)
- Fidel Castro overthrows General Batista in Cuba (1959)
- US Federal Government forces Arkansas to desegregate its schools (1959)

- The International Geophysical Year – actually 18 months from July 1957 to December 1958 – marks a turning-point in polar exploration as 67 nations collaborate in a joint research project setting up bases in both the Arctic and Antarctic

- The USSR launches Sputnik I, the first satellite to orbit Earth, in October 1958; a month later it is followed by Sputnik II, with the dog Laika on board

- Superglue, the first cyanoacrylate-based adhesive, is launched

- While working for Texas Instruments, Jack Kilby develops the integrated circuit

- Brussels hosts Expo 58, the first major World's Fair since before World War II

- Joseph-Armand Bombardier, a Canadian businessman, develops the snowmobile

- Working for Volvo, Sweden's Nils Bohlin invents the three-point safety belt

1960

- Racial strife in South Africa leads to the Sharpeville massacre (1960)
- Decolonisation speeds up in Africa as Nigeria gains independence from Britain and most of France's possessions win self-rule (1960)
- The Berlin Wall is constructed by the East German authorities (1961)
- US-backed forces launch a failed invasion of Cuba at the Bay of Pigs (1961)

- Calor launches the rigid-hood hair dryer

- Theodore H Maiman creates the laser beam

- NASA launches the first communications satellite into space

- The 11th General Conference on Weights and Measures establishes the International System of Units, standardising the metric system

- The audio cassette, put on the market by Philips, challenges the supremacy of vinyl records

- The first airtight milk cartons appear, made by Tetra Pak

- In the USA, Charles Koester and Charles Campbell pioneer laser eye surgery techniques

- Unimate, built by a firm called Unimation, joins the General Motors assembly line, the first use of robotics in car manufacture

- Russian Yuri Gagarin becomes the first person to travel in space, making an orbit of the Earth on board the Vostok 3KA-3 capsule; a month later Alan Shepard of the USA became the second man in space

- The Digital Equipment Corporation produces the PDP-1, the first interactive computer

▶ PDP-8 computer

▼ Unimate robot

▲ Yuri Gagarin makes headlines

CHRONOLOGY

1962

EVENTS
- The Cuban Missile Crisis is brought to an end when the USSR withdraws its nuclear missiles from Cuba (1962)
- A Moscow-Washington hotline is set up to help defuse future crises following the signing of the nuclear Test Ban Treaty (1963)
- US President John F Kennedy is assassinated in Dallas (1963)

INVENTIONS
- The introduction of Data-Phone modems, precursors of the digital ADSL modem, makes it possible to transfer data between computers over telephone lines
- Australia's John Dickenson invents the modern hang glider
- William Buehler takes out a patent in the USA on Nitinol, an alloy of nickel and titanium, starting the development of shape memory alloys
- Valentina Tereshkova of the Soviet Union becomes the first woman to journey into space
- IBM launches the BASIC computer programming language
- The French firm Elco-Lesieur takes out a patent on plastic bottles

1964

EVENTS
- Congress passes the Civil Rights Act, making racial discrimination illegal in the USA (1964)
- Martin Luther King wins the Nobel peace prize (1964)
- The US military launches air strikes on North Vietnam (1965)
- Ian Smith declares UDI – a Unilateral Declaration of Independence – in Southern Rhodesia, now Zimbabwe (1965)

INVENTIONS
- Robert Moog invents the Moog synthesiser, theoretically capable of producing an infinite variety of sounds
- Patented by the Ministry of Defence, carbon fibre is launched on the market as a new, ultra-resistant synthetic material
- A German affiliate of IBM develops word processing as a means of handling its internal paperwork
- US paint manufacturer Reeves markets the first acrylic paints
- The US glass manufacturing company Corning patents photochromic lenses
- Mary Quant launches the mini-skirt, which soon becomes a worldwide fashion phenomenon
- Alexei Leonov of the Soviet Union makes the first space walk
- Intelsat 1 launches the era of satellite communications

▶ Mini-dresses designed by Mary Quant

▼ Valentina Tereshkova

◀ Close-up of a liquid-crystal display screen

CHRONOLOGY

1966

- The Cultural Revolution gets under way in China (1966)
- Israel defeats Egypt in the Six Days War (1967)
- Student riots disrupt France (1968)
- Soviet tanks crush the Prague Spring in Czechoslovakia (1968)
- Martin Luther King is assassinated (1968)
- Tet offensive strikes at Saigon in Vietnam (1968)

- The Rance tidal power station opens in Brittany
- In France the Aérotrain is tested, but never put into service
- The introduction of container ships transforms international commerce
- Working at a Cambridge radio astronomy observatory, Antony Hewish and Jocelyn Bell identify pulsars
- The first windsurfing sailboards appear on the US West Coast
- James Drake improves sailboard design by adapting the wishbone boom as a means of controlling the sail
- Working at the Stanford Research Institute, Douglas Engelbart invents the computer mouse
- RCA Laboratories in the USA present the results of research done by George Heilmeier and his team in developing liquid-crystal display (LCD) technology – liquid-crystal screens follow shortly after
- In England, Giles Brindley does pioneering work on providing blind patients with some sensation of sight

1969

- Willy Brandt becomes chancellor of West Germany and launches his Ostpolitik policy of closer ties with East Germany (1969)
- In Libya, Colonel Muammar al-Gaddafi overthrows the monarchy and proclaims the Libyan Arab Republic (1969)
- British troops are sent into Northern Ireland (1969)
- Anwar El Sadat becomes president of Egypt (1970)

- IBM launches the first computer floppy disk
- The US physicist Murray Gell-Mann wins the Nobel prize for physics for the discovery of quarks, elementary particles of matter
- The Apollo 11 mission successfully lands on the Moon and Neil Armstrong becomes the first human to walk there
- The ioniser is invented
- Working at Stoke Mandeville Hospital, Rog Maling and Derek Clarkson create the Patient-Operated Selector Mechanism (POSM), a device enabling severely handicapped people to perform some functions for themselves and to communicate with others
- In Baltimore, Michel Mirowski and Morton Mower create a prototype implantable defibrillator

▼ The Apollo 11 crew – left to right: Neil Armstrong, Michael Collins and Buzz Aldrin

▼ Container ships first appeared in the 1950s and rapidly revolutionised shipping

CHRONOLOGY

1971

EVENTS
- General Idi Amin seizes power in Uganda in a military coup
- China is admitted to the United Nations General Assembly
- Bangladesh secedes from Pakistan to become an independent nation
- Australia and New Zealand withdraw troops from Vietnam

INVENTIONS
- The US firm of Alton Box Board develops an effective water jet cutter
- Marcian Hoff, a US engineer working for Intel, develops the Intel 4004, the first microprocessor
- The Soviet Union puts Salyut 1 into orbit, the first space station designed for long-stay manned flights
- Donald Wetzel, a US engineer, applies for a patent on an automatic teller machine (ATM)
- Bausch & Lomb, a US optical firm, markets the first soft contact lenses
- Inspired by surfboards, snowboards and monoskis start to appear on ski slopes
- US journalist Don Hoefler publishes a series of articles in *Electronic News* under the general heading 'Silicon Valley, USA' – the term sticks for the technology hub developing on the outskirts of San Francisco

1972

EVENTS
- The Watergate building is bugged and the political scandal involving President Nixon gets under way in the USA
- 14 civil-rights protesters in Londonderry are killed by the British army in what becomes known as Bloody Sunday
- Palestinian terrorists stalk the Munich Olympic Games

INVENTIONS
- François Gernelle, working for the French firm R2E, develops the Micral N, the first microcomputer
- In the USA, Texas Instruments puts the first pocket calculator, the TI-2500, on the market
- Working from ideas put forward by the physicist Allan Cormack, Godfrey Hounsfield and Jamie Ambrose developed a working CT scanner for use on humans
- Edward Lorenz, a US mathematician and meteorologist, expounds chaos theory
- The Airbus A300, the first twin-engined widebodied jet airliner, takes to the skies

▶ Automatic teller machine (ATM) – better known as a cash point

▼ Water jet cutter

▼ Snowboard in action

150

CHRONOLOGY

1973

- In Chile, General Pinochet leads a military coup that overthrows the socialist government of President Salvador Allende
- The Yom Kippur War pits Israel against neighbouring Arab states
- The Paris Peace Accords are signed, ending direct US involvement in the Vietnam War

- The US chemist Paul Lauterbur produces the first images obtained by nuclear magnetic resonance imaging
- IBM launches the 3340, the first commercial hard disk drive for computers
- In the USA, skateboards become popular after Frank Nasworthy replaces the roller-skate wheels previously used with polyurethane wheels
- Working with a firm called Teldix, the German company Bosch develops an anti-lock braking system (ABS) for motor vehicles
- NASA launches the first US space station, Skylab

1974

- President Juan Perón of Argentina dies, and is succeeded by his wife Isabel
- Portugal's Carnation Revolution sets the nation on the path to democracy

- Californian trail-bike enthusiasts popularise the mountain bike, originally devised by Geoff Apps in England six years earlier
- In Ethiopia the discovery of the remains of Lucy, an early hominid, extend scientists' understanding of the evolution of the human race
- Erno Rubik, a Hungarian professor of architecture, devises his cube
- In the USA, General Motors introduce catalytic converters to reduce vehicle exhaust pollution
- France's Roland Moreno invents the smart card
- US astronomers Russell A Hulse and Joseph H Taylor observe the first binary pulsar, located in the Aquila constellation

▼ Mountain bikes take off

▼ Lucy, an early human ancestor

◄ Rubik's Cube

CHRONOLOGY

1975

EVENTS
- In Spain, General Franco dies, and is succeeded by King Juan Carlos I (1975)
- The Khmer Rouge come to power in Cambodia following the capture of Phnom Penh (1975-9)
- In China Mao Zedong and Zhou Enlai both die (1976)

INVENTIONS
- Working at the University of Aberdeen, John Hughes and Hans Kosterlitz isolate endorphins (which they called 'enkephalins'), publishing their findings in the journal *Nature*
- Bic launches the disposable razor, another milestone in the development of the throw-away society
- The French vulcanologist Haroun Tazieff travels around the world studying volcanoes that he will later make the subject of books and films introducing vulcanology to the public
- Andreas Pavel of Germany takes out patents on his Stereobelt personal stereo, two years before Sony launch the Walkman
- The Konica C35 AF goes on the market as the first camera equipped with autofocus
- The German cardiologist Andreas Gruentzig develops the technique of coronary angioplasty, seeking to prevent heart attacks by dilating narrowed arteries
- In the USA Bill Gates and Paul Allen found Microsoft

1978

EVENTS
- Egypt and Israel sign the Camp David Accords
- The European Monetary System (EMS) is established, but Britain declines to join
- Democratic Republic of Afghanistan is proclaimed
- Black majority rule agreed in Rhodesia (Zimbabwe)

INVENTIONS
- The Argos system is established to collect environmental data worldwide, locating the source via geo-positioning satellites
- Louise Brown, the first baby conceived through in-vitro fertilisation (IVF), is born in Oldham, Greater Manchester
- The use of optical character recognition and text-to-speech synthesis devices spreads in the USA thanks to the work of Raymond Kurzweil
- The first remote-control device for locking and unlocking car doors is developed; it becomes known as a 'plip key' after its inventor, Paul Lipschutz

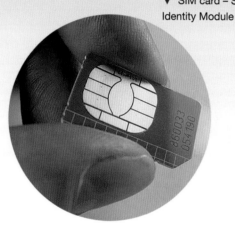

▼ SIM card – Subscriber Identity Module

▼ Haroun Tazieff

▲ Pulsar

CHRONOLOGY

1979

- In Iran, the Shah is driven into exile by an Islamic revolution led by Ayatollah Khomeini (1979)
- The USSR invades Afghanistan at the request of the country's pro-Soviet government, starting a nine-year war (1979)
- Iran is invaded by its neighbour, Iraq (1980)

- The Canadians Scott Abbott and Chris Haney invent Trivial Pursuit
- Scott and Brennan Olson put inline wheels on ice-hockey boots, subsequently merchandising the skates through their Rollerblade company
- The US astronomers Peter Young and Wallace Sargent offer proof of the existence of a black hole within a galaxy
- Sony and Philips both start marketing compact discs
- The lithotriptor, which uses acoustic shock waves to destroy kidney stones, is invented
- An engineer named Bill Carlton devises the Cyclops 'magic eye' system of infra-red beams used at Wimbledon to adjudicate on tight line calls

1981

- General Jaruzelski declares martial law in Poland; the Solidarity trade union – established the previous year – is dissolved (1981)
- Hissène Habré seizes power in Chad (1982)

- In the USA the 3M Company patents Post-It notes
- Paul MacCready, a US aeronautical engineer, develops an aircraft powered by solar energy
- Adam Osborne designs the first commercially available portable computer
- Working for IBM's Zurich research group, Gerd Binning and Heinrich Rohrer invent the scanning tunnelling microscope
- The French biologist François Jacob, joint winner with Jacques Monod and André Lwoff of the 1965 Nobel prize for medicine, publishes *The Possible and the Actual*

▲ André Lvoff, François Jacob and Jacques Monod set off to collect their Nobel prize

◀ A microscope image of beta-endorphins

Index

Page numbers in *italics* refer to captions.

A
A4 rocket 72-3
Abbott, Scott 153
acrylic paints 54, *54*
active-matrix technology 65-6
Aérotrain 58, *58*
AIDS 142
air traffic control 85, *86*
air-cushion technology 58
Airbus A300 83, *83*, 84, 85
Airbus A380 48, *49*
airliners 82-3
   Airbus A300 83, *83*, 84, 85
   Airbus A380 48, *49*
   Boeing 707 82
   Boeing 747 83, 84, 85, *86*
   Boeing 787 48
   Caravelle 82
   Comet 82
   Concorde 84
   Convair 990 82
   Douglas DC-8 82, *83*
   Douglas DC-10 82
   Lockheed Tristar 82, 84
   McDonnell-Douglas DC-10 84
   solar aircraft 153
   Tupolev 82, *82*
   Vickers VC10 82
   see also civil aviation
Akkersdijk, Erik 125
Aldrin, Edwin 'Buzz' 74, 76, 77, *77*, 78
Allen, Paul 103
alloys
   carbon–carbon alloys 48
   shape memory alloys 42, *42*
Altair 8800 computer 103
Alton Box Board 87
aluminium cans 109, *109*
Ambrose, Jamie 116
Amdahl, Gene 105
Ames Research Center 143
Amstrad PCWs 104
answering machines 29
anti-lock braking 119
Apollo programme 72, 74-9, 113
Apple computers 52, *52*, 89, 104, *104*, 106
Apps, Geoff 124
Ardipithecus ramidus 134
argon lasers 30
Ariane rockets *24*
Arimura, Kunitaka 126
Armistead, W C 54
Armstrong, Neil 74, 76, 77, 78
asbestos 46, 49

ASCII 103
Asterix satellite 23, *23*
astronauts
   physical and psychological problems 92-3
   qualities 91
   survival training 91
   see also individual astronauts
astrophysics 60-1
AT&T 24, 26
Atlas-Centaur rockets 23
ATMs 98, *98*
atomic bombs 22
atomic clocks 146
ATOS-VR hang gliders 35
audio cassettes 29, *29*, 130
Australopithecus afarensis 134-5
Avdeiev, Sergei 95

B
BAC One-Eleven 82
Bacon, Roger 46
Baez, Joan *140*
Bakelite 46
ballpoint pens 108, 109
Barbie doll® 146
Barish, Dave 35
BASIC programming language 105
batteries, rechargeable 111
BBC Micro 104
Beat Generation 141, 142
Beatles 43
Beck, Claude 81
Becquerel, Antoine 26
Bell, Jocelyn 60, 61
Bell Laboratories 31
Belyayev, Pavel 91
Bennett, Bill 34
Bentz, Melitta 108
Berner, Bob 103
Bertin, Jean 58
Bic Crystal 109
Bic Phone 109
Bich, Marcel 109
Bichet, Pierre 44
Big Brother and the Holding Company 142
Binning, Gerd 153
biodegradable products 109, 111
biometric passports 129
bioplastics 109
BIOS (Basic Input/Output System) 106

bird flu 86
Biro, Lazlo and Georg 108, 109
black holes 153
Bloch, Felix 117
BMX (bicycle motocross) racing 124
Boardman, Chris 47
boat building 48
Boeing 707 82
Boeing 747 *83*, 84, 85, *86*
Boeing 787 48
Boeing KC-135 Stratotanker 93
Bohlin, Nils 147
Bombardier, Joseph-Armand 147
bonsai trees 131
Bosch 119
bosons 71
Braess, Dietrich 121
Branly, Édouard 27
Braun, Wernher von 22, 26, 72-3, *72*, *73*, 75
Bravo program 52
Brindley, Giles 68
Brown, Louise 152
Buehler, William 42
Burroughs, William 141

C
cable cars *140*
cable modems *31*
calculators 64, 109, *111*, 112-13, *112*, *113*, 130
Calor hair dryer 147
cameras
   autofocus 152
   disposable 109, *111*
   mini-cameras *131*
Campbell, Charles 30
Capek, Karel 36
Caravelle jetliners 82
carbon fibre 46-9, *46*, *47*
carbon–carbon alloys 48
cardioverters 81
Carlos, Walter 43
Carlton, Bill 153
Carpenter, Jake Burton 100
cars 120-3
   accidents 122
   anti-lock braking 119
   car-sharing 122
   catalytic converters 125
   electrically powered cars 123, *123*
   hybrid cars 123
   ownership 121-2
   remote-control locking 152
   robotic vehicle assembly 40, *40*
   safety belts 147

traffic congestion 120-1, *120*, *121*
traffic-reduction schemes 122-3
cash machines 98, *98*
cassette players 29
Castrucci, Paul 126
catalytic converters 125
cathode-ray tubes 66
cell behaviour 50-1
cell phones 132
cellulose 108
centrifuge 93
Cernan, Gene 76, 79
Chaffee, Roger 76
Challenger space shuttle 91, 92
chaos theory 114-15, *114*, *115*
charter flights 85
chip 'n' pin cards 129
Chowning, John 43
Christiansen, Ole Kirk 146
cigarette lighters 108
Cisco 139
civil aviation 82-6
   air traffic control 85, *86*
   charter flights 85
   freight transport 85
   inflight service *83*, 85
   low-cost airlines 85
   security issues 85-6
   see also airliners
civil rights movement *140*
Clarkson, Derek 80
Coca-Cola 92, 108-9
Cockerell, Christopher 58, 146
coffee filters 108
coherers 27
Cold War 20, 74, 130
   see also space programme
Collins, Michael 74, *76*
Columbia command module 78, *78*
Columbia space shuttle 92
Comet jetliners 82
Commodore PET computer 104
compact discs 29
composite materials 47, 48, 78, 87
computers
   Apple 52, *52*, 53, *104*, 104, 106
   ASCII 103
   computer-assisted machine tools 146
   flash cards 118
   floppy disks 69, *69*
   hard disk drives 118, *118*, *132*
   laptops 63, 66, 131
   microdrives 118
   microprocessors 27, 40, 88-9, *88*, *89*, 130, 132

Microsoft 52-3, 103, 104
modems 31, *31*
mouse 63, *63*
operating systems 105
PCs 53, 89, 102-7, *103*
programming languages 105
PROM memory 126
RAM memory 105, 106, 118
ROM memory 106
smart cards 126-9, *126, 128, 129*
spreadsheets 53
trackballs 63
word processing 52-3, *52, 53*
Concorde 84
congestion charging 122-3, *122*
contact lenses 99, *99,* 109
container ships 59, *59*
contraceptive pills 146
Convair 990 82
Coppens, Yves 134, *135*
Cormack, Allan 116
Corning 54
coronary angioplasty 152
counterculture 142
Courier 1-B satellite 20
Courrèges, André 55
CP/M operating system 105
CT scanning 116, *116*
cybernetics 37
cycle lanes 122, *123*
Cyclops 'magic eye' system *153*

D
Dalton, R H 54
Damadian, Raymond 117, *117*
Damblanc, Louis 26
Darby, Newman 62
data security 129
Dataphones 31
De Forest, Lee 27
deep fat fryers 42
defibrillators, implantable 81, *81*
Dellinger, David *53*
Dethloff, Jürgen 126
Deutsch, Herbert 43
Devol, George 37-8
Dezhurov, Vladimir 96
Diamant rockets 23
diamonds, synthetic 146
Dickenson, John 32
Dickson, James 108
Dictaphones 29
digital cassettes 29
digital modems 31
DIHEL Alphatronic 102
diodes 27
disposable products 108-11
DNA 51
Dobelle, William 68
Dolby system 29

DOS operating system 105
dot.com bubble 139
Douglas DC-8 82, *83*
Douglas DC-10 82
Doyle, Mike 100
Drake, James 62
drive-in vaccination clinic *121*
DSL (digital subscriber line) modems 31
Duralumin 46
Duret, Xavier 101
Dylan, Bob *140*

E
Early Bird satellites 24
East London Tech City 139
Eastman Kodak 138
EasyJet 85
Edison, Thomas 46
El Niño *24*
electromagnetic waves 27
electronic eyes 68, *68*
electronic money *see* smart cards
Elektro (robot) 37
endorphins 136-7, *137*
Engelbart, Douglas 63
Engelberger, Joseph 38
ENIAC computer 132
epoxy resins 47
Essen, Louis 146
Ethernet 138
Excel 53
excimer lasers 30
Exploratorium 143
Explorer 1 satellite 21, *22*
Expo 58 147
Eyjafjallajökull eruption 86

F
Fairchild Semiconductors 138, *138*
Famulus robotic arm 39
fax machines 31
Fergason, James 64
fermions 71
Feynman, Richard *133*
Fillmore West *141*
films, robots in 37, *41*
flash cards 118
Fleming, John Ambrose 27
flight simulators *84*
floppy discs 69, *69*
Flower Power 142
Focal computer language 103
food packaging 108
Fortran programming language 105
Franz, Norman 87
Fraze, Erman 108

Freedom space station 94, 96-7
freight transport
 air freight 85
 container ships 59, *59*
Frenkel, Michel 125
Fuller, Calvin 26
functional MRI (fMRI) 117

G
Gagarin, Yuri 75, 90-1, *90*
games and toys
 Barbie doll® 146
 Lego 146
 Rubik's Cube 125, *125*
 Trivial Pursuit 147
Gates, Bill 103, 105
gay community 142, *142*
Gebreselassie, Haile 136
Gell-Mann, Murray 70-1
Gemini 4 mission *91*
General Electric 46, 138
General Motors 125
genes 50-1
geolocation 24, 25
geostationary orbit 21, 24
Gernelle, François 102, 103
Gibson, Edward *95*
Gibson, Robert 96
Ginsberg, Allen 141, 142, *142*
glass fibre 46
Glenn, John 75
GLOSNASS satellite system 25
Glushko, Valentin 75
go-karts 146
Goddard, Robert 26
golf clubs 48
GPS 25
Graham, Bill *141*
Granqvist, Eirik *135*
Grateful Dead 142
Gray, Tom 134
green technology 139
Greyhawk Systems 66
Grissom, Gus 76
Grotrupp, Helmut 126
Gruentzig, Andreas 152
Grumman Aircraft Corporation 76
GUI (graphical user interface) 106
Guignard, André 63
Guillemin, Roger 136
gyrocopters 34

H
Haas, Earle 108
Hadfield, Chris *39*
Haight-Ashbury 142
hair dryers 147
Halse, Fred 78
Haney, Chris 153

hang gliders 32-5, *33, 35*
hard disc drives 118, *118, 132*
Hawk, Tony 119
Heilmeier, George 64
Helios satellites *24, 26*
HEMA hydrogel 99
Herbert Televox (robot) 36, *37*
Hero of Alexandria 98
Hertz, Heinrich 27
Hewish, Antony 60, *60,* 61
Hewlett, William *138*
Hewlett Packard *138,* 139
hippies 142
Hoefler, Don 138
Hoff, Marcian 88
Holter monitors 81
home furnishing 111
hominid fossils 134-5, *134, 135*
Horizons-2 satellite *25*
Hounsfield, Godfrey 116
Hounsfield scale 116
hovercraft 146
hovertrains 58, *58*
HP-65 pocket calculator *112, 113*
Hubble space telescope 24
Hughes, John 136
Hulse, Russell Alan 61
HX 20 laptop 66
hybrid cars 123

I
IBM 69, 89, 103, 118
IBM 5100 computer 103
IBM operating system 104
ICDs (implantable cardioverter-defibrillators) 81, *81*
iMac G3 computer *107*
information and communication society 107 *see also* computers
Ingels, Art 146
Inglessi, Jean 109
integrated circuits 27, 88, 130
Intel 88, 104
Intel 4004 microprocessor 88, *88*
Intelsat 1 satellites 24
International Geophysical Year 147
International Space Station (ISS) 21, *39,* 96-7, *97*
International System of Units (SI) 28
Interntional Civil Aviation Organisation (ICAO) 83
Iomega 69
ionisers 80, *80*
ionising radiation 116, 117
Ive, Jonathan *107*
IVF (in-vitro fertilisation) 152

INDEX

J
Jacob, François 50-1, *50, 51*
Jacquard, Joseph 36
Jalbert, Domina 35
Jarre, Jean-Michel 43
Jefferson Airplane 142
Jobs, Steve 63, 103-4, 138
Johanson, Donald 134
Johnson, Reynold 118
Joplin, Janis 142
jukeboxes 28, *28*

K
Kang, Andrew 125
Kapany, Narinder 146
Kaplan, Victor 56
Kennedy, John F 72, 74-5
Kerouac, Jack 141
kidneys, artificial 146
Kilby, Jack 27, 112, 130
kitesurfing 62
Kleenex paper hankies 108
Koester, Charles 30
Kolff, William 146
Korolev, Sergei 75
Kosterlitz, Hans 136
Kraftwerk *43*
Kuka 39
Kurzweil, Raymond 152

L
lac operon mechanism 50
Laika 21, *21*
laptops 63, 66, 131
laser eye surgery 30, *30*
laser photocoagulation 30
laser printers 138
LASIK(Laser-Assisted In-situ Keratomileusis) 30, *30*
launch vehicles *see* rockets
Lauterbur, Paul 117
laws of motion 26
LCD (liquid crystal display) technology 64-7
Leary, Timothy 140-1
LEDs (light-emitting diodes) 63, 67
Lego 146
Lehmann, Otto 64
Lelouch, Claude 28
Leonov, Alexei 91-2, *91*, 148
Lilienthal, Otto 32
Linux operating system *105*, 107
Lipschutz, Paul 152
Liquitex 54
lithotriptors 153
Liwei, Yang 93
Lockheed Tristar 82, 84
long-life milk cartons 29

Lorenz, Edward 114-15
Lovell, Jim 78, *78*
LSD 141, 142
Lucy 134-5, *135*
Lukasiewicz, Jan 112
Luna 7 39
lunar rovers 39, 79
Lunboikodh 1 robot 39
Lutus, Paul 52
Lwoff, André 50, *51*

M
McCandless, Bruce 92
McDonnell-Douglas DC-10 84
machine tools 146
Macintosh Classic *106*
Macintosh Lisa 63, *104*, 106
McLean, Malcolm 59
McLellan, William *133*
Macready, Paul 153
MacWrite 52
Maglev trains 58
magnetic resonance (NMR) techniques 117
Maiman, Theodore 30
Maling, Reg 80
Mandelbrot, Benoit 114
Manned Maneuvering Unit 92
Mansfield, Peter 117
Mars missions 92, 93, 97
Mars rovers 39
Marshall Space Flight Center 73
Mathieu, Frédéric 28
Mattel 146
matter, particles of 70-1
medicine
  contraceptive pill 146
  coronary angioplasty 152
  CT scanning 116, *116*
  defibrillators, implantable 81, *81*
  IVF (in-vitro fertilisation) 152
  kidneys, artificial 146
  laser eye surgery 30, *30*
  lithotripter treatment 153
  microsurgery 133, *133*
  MRI scanning 117, *117*
  neuroendocrinology 136-7
  prosthetics 48, 133
Mellotron 43
memory cards 118, *118*, 126, 128
Mercury spacecraft 74, *75*
messenger RNA 50, *51*
Mestral, George de 146
meteorology 24-5
  chaos theory and 114
Micral computer 102, *103*
microcassettes 29
microchips *see* microprocessors
microcopters 132, *132*

microdrives 118
microlights 34, *34*
microprocessor cards 126-7
microprocessors 27, 40, 88-9, *88, 89*, 130, 132
microprosthetics 133
Microsoft 52-3, 103, 104
Microsoft Office 53
microsurgery 133, *133*
Milacron Corporation 39
Milk, Harvey 142, *142*
Mills, Victor 146
Milovich, Dimitri 100
mini-skirts 55, *55*
miniaturisation 27, 81, 88, 130-3
Mir space station 23, 95-6, *95*
Mirowski, Michel 81
mobile phones 132
modems 31, *31*
Monod, Jacques 50-1, *50, 51*
monoskis 100-1, *101*
Moody, Helen Wills 55
Moog, Robert 43
Moog synthesisers 43, *43*
Moon missions
  Apollo programme 72, 74-9, 113
  lunar rovers 39, 79
  Moon landings 77, 78, 79
Moon rocks 78, 79
Moore, Gordon 89
Moreno, Roland 126, 127, *127*
morphine 136, 137
Moscone, George 142
Motorola 89
Mount Etna 25, *45*
mountain bikes 123, *124*
mouse (computer) 63, *63*
Mower, Morton 81
Moyes, Bill 34
MRI scanning 117, *117*
MS-DOS operating system 105
music
  cassettes 29
  compact discs 29
  jukeboxes 28, *28*
  synthesisers 43, *43*

N
Naish, Robby 62
Nakajima, Yu 125
nanotechnology 130-3
nappies, disposable 108, *108*, 111, 146
NASA (National Aeronautics and Space Administration) 20, 24, 32, 33, 35, 73, 75, 78, 143
Nasworthy, Frank 119
Naviplane 58
Ne'eman, Yuval 70

Neoprene 46
neuroendocrinology 136-7
neutron stars 60-1
Newton, Sir Isaac 26
Nicoud, Jean-Daniel 63
Nitinol 42
North American Aviation 76

O
Oberth, Hermann 72
oil crisis (1970s) 122
Ölander, Arne 42
Olson, Scott and Brennan 153
Olympus 29
OOP (object-oriented programming) 105
ophthalmology
  contact lenses 99, *99*, 109
  laser eye surgery 30, *30*
  photochromic lenses 54
  retinas, artificial 68, *68*
Oppenheimer, Frank 143
optical character recognition 152
optical fibres 146
Oracle *139*
orbits 21
origins of humankind 134-5
Orrorin tugenensis 134
orthodontics 42
Osborne, Adam 153
Oyster cards 128

P
P-1 paraplane 35
pacemakers 81
packaging 108, 111
Packard, David *138*
pain, experience of 137
palaeontology 134-5
Palmer, Barry Hill 32
Pampers nappies 146
Pao, Michael 87
paper hankies 108
paper towels 108
paragliders 35, *35*
Parasev (Paraglider Research Vehicle) 33
PARC (Palo Alto Research Center) 138
particle accelerators 71
particle physics 70-1
Pavel, Andreas 131
PC-DOS operating system 105
PCs (personal computers) 53, 89, 102-7, *103*
PDP-1 computer 103
PDP-8 computer *102*, 103
Pearson, Gerald 26
Permanent Pigments 54
personal stereos 131

INDEX

PET (polyethylene terephthalate) 108
pets, robot 41
Philips 29
photochromic lenses 54
photoelectricity 26
photorefractive keratectomy (PRK) 30
photosensors 42
photovoltaic cells 22, 94, 113
Pickering, William 22
Pincus, Gregory G 146
Pink Floyd 43
pixels 64, 65, 66
plasma technology 67
plastic bottles 148
plastics 108-9, 111
PLEDs (polymer light-emitting diodes) 67
Plexiglas 99
'plip keys' 152
polar exploration 147
Poliakov, Valeri 95
Pollard, Willard 37
polyacrylonitrile 46
polyester 47
polymers 46
Popov, Alexander 27
Poppen, Sherman 100
POSM (Patient-Operated Selector Mechanism) 80, 80
Post-It notes 153
potassium–argon dating 135
Powerpoint 53
PROM (Programmable Read-Only Memory) 126
prosthetics 48, 133
Proton rockets 75
PSTN (Public Switched Telephone Network) modems 31
psychedelic revolution 141
public transport 123
  see also civil aviation
pulsars 60-1, 61
PUMA (Programmable Universal Manipulation Arm) 39
Purcell, Edward Mills 117

Q
Quant, Mary 55
quarks 70-1, 70
quasars 60, 61

R
R7 missile 20
racing cars 48
radio telescopes 60
RAM (Random Access Memory) 105, 106, 118
Ramac 305 hard disk drive 118, 118
Rance tidal power station 56-7, 57
Rausing, Ruben 29
Raysse, Martial 54
razors, disposable 109
RCA laboratories 64
recycling 110, 110, 111
Redstone missile 21, 73, 73
reductionism 131
Reinitzer, Friedrich 64
remote-control locking 152
retinas, artificial 68, 68
reverse-thrust mechanism 58
RFID (radio frequency identification) 128
Richard, Bill and Marc 119
Richards, Charles 33
Ride, Sally 91
ring-pull cans 108, 109
Rivera, Diego 54
RoboCup 40
robotics 36-9
  car manufacture 40, 40
  military robots 41
  nuclear-powered robots 39
  robot pets 41
  robotic arms 37, 38-9, 39
  space robots 39
Rock, Arthur 138
Rocket City (Huntsville, Alabama) 73
rockets
  A4 72-3
  Ariane 24
  Atlas-Centaur 23
  Diamant 23
  Proton 75
  R7 20
  Saturn 73, 74, 75, 75, 76
  Semyorka 23, 75
  Thor Delta 23, 26
  Titan 23
  Vanguard TV-3 21
Rogallo, Francis 32, 32
Rohrer, Heinrich 153
Rollerblades 153
Rolling Stones 43
ROM (Read Only Memory) 106
Roomba vacuum cleaner 41
Roselund, Harold 37
Rothko, Mark 54
RPN (reverse Polish notation) 112, 113
Rubik, Erno 125
Rubik's Cube 125, 125
Rubinstein, Seymour 52
Ruelle, David 114
Ryanair 85
Ryle, Martin 61

S
SAFER (Simplified Aid for EVA Rescue) 92
safety belts 147
safety valves 42
Sahelanthropus tchadensis 134
sailboards 62, 62
Salyut space stations 23, 91, 94
SAMOS satellites 24
sampling 43
San Andreas Fault 141
San Francisco 140-3
sanitary towels 108
Sargent, Wallace 153
SARS epidemic 86
satellites
  anatomy of a satellite 22
  communications satellites 20-7, 73
  electronics 27
  mini-satellites 27
  reconnaissance satellites 24, 24, 26
  weather satellites 24, 24, 27
Saturn V rocket 73, 74, 75, 76
scanning tunnelling microscopes 153
Schalamon, Wesley 46
Schally, Andrew 136
Schmitt, Harrison 79
science museums 143
Scopitone 28, 28
security, airport 85-6
SED (surface-conduction electron-emitter display) 67
Seiko Epson 66
Semyorka rockets 23, 75
shape memory alloys 42, 42
Sharkey (robot) 38, 38
Sharp 64, 67
Shenzhou 5 spacecraft 93
Shepard, Alan 75, 91
Shepherd-Barron, John 98
Shindo, Akio 46
Shockley, William 27, 138
shopping bags, plastic 109
Silicon Fen 139
Silicon Glen 139
Silicon Valley 88, 138-9, 139, 143
SIM cards 129
Simjian, Luther 98
Simon, Nathalie 62
Simonyi, Charles 52
Sinclair ZX81 104
Siquieros, David Alfaro 54
skateboards 119, 119
skiing 101
skwaling 101
Skylab 94, 95
skytrikes 34
slope soaring 35
smart alloys 42
smart cards 126-9, 126, 128, 129
smart fabrics 42
Smith, Willoughby 26
snowboards 100-1, 100
snowkiting 101
snowmobiles 147
Snyder, Steve 35
Sojourner rover 39
solar aircraft 153
solar panels 26, 26
Sony 41, 131
Souviron, Gil 35
Soyuz spacecraft 94, 94, 97
space debris 23, 23
space probes 39
space programme
  animals in space 21, 21
  Apollo programme 72, 74-9, 113
  astronaut qualities 91
  disasters 76, 92
  first manned flight 91
  Gemini 4 mission 91
  Mars missions 92, 93, 97
  robotics 39
  satellites 20-7, 73
space shuttles 48, 49, 91, 92
space stations 21, 23, 93, 94-7, 94, 95
  Freedom 94, 96-7
  International Space Station (ISS) 21, 39, 96-7, 97
  Mir 23, 95-6, 95
  Salyut 23, 91, 94
  scientific experiments 97
  Skylab 94, 95
space tourism 97
space walks 91-2, 91, 95, 148
spacesuits 48, 91, 92
sports
  BMX (bicycle motocross) racing 124
  Formula 1 racing 48
  golf 48
  hang gliding 32-5, 33, 35
  kitesurfing 62
  monoskiing 100-1, 101
  mountain biking 123, 124
  paragliding 35, 35
  Rollerblades 153
  skateboarding 119, 119
  skiing 101
  skwaling 101
  snowboarding 100-1, 100
  snowkiting 101
  windsurfing 62, 62
SPOT satellite 25
spreadsheets 53
Sputnik 1 20, 21, 27
Sputnik 2 21
spy planes 22
Stafford, Tom 76

INDEX

stainless steel 46
Stanford Industrial Park 138
Stanford University 142-3
Stereobelt 131
Stookey, S D 54
'strange attractor' phenomenon 115
Summer of Love 141-2
Superglue 147
Swigert, Jack 78, *78*
swine flu 86
synchotrons 70
synthesisers 43, *43*

T

T3 Tomorrow Tool 39
Taïeb, Maurice 134
Takens, Floris 114
tampons 108
Tandy Radio Shack 104, *106*
tape recorders 130
Taylor, Joseph Hooton, Jr 61
Tazieff, Haroun 44-5
Teflon 46
telephones
  Bic Phone 109
  cell phones 132
  phonecards 126, *126*, 128
television 130, *131*
  LCD screens 66-7, *67*
Telstar satellites 24
Terraplane 58
Teseshkova, Valentina 91, 148
Tetra Paks 29
Texas Instruments 112
text-to-speech synthesis 152
TFT (thin film transistor) system 65
TGV (*Train à Grande Vitesse*) 58
Thagard, Norman 96
Thor Delta launch vehicle 23, *26*
3M Company 153
Three Mile Island nuclear plant 41
tidal power stations 56-7, *57*
tights 55
TIROS satellites 24
Titan rockets 23
Tito, Dennis 97
TOPEX-Poseidon satellite *24, 26*

Torvalds, Linus 107
Toumaï fossil 134, *134*, 135
Townshend, Brent 31
trackballs 63
traffic congestion 120-1, *120, 121*
traffic pollution 122
traffic reduction schemes 122-3
trams *123*
Transamerica Pyramid 143, *143*
transistor radios 130, 131
transistors 27, 88, 89, 138
TRANSIT satellite system 25
Transrapid hovertrains 59
triodes 27
Trivial Pursuit 147
TRS-80 computer 104
Tsiolkovsky, Konstantin 26
Tuohy, Kevin 99
Tupolev jetliners 82, *82*
turbines 56, *56*
Turing, Alan 27
Twiggy 55, *55*

U

U-2 spy plane 22
Ugon, Michel 126-7
Unicode 103
Unified Modelling Language 105
Unimate (robot) 36, *36*, 38
Unimation 38
Union Carbide 46
Unix operating system 105
unleaded petrol 122

V

V-1 and V-2 missiles 26, 72, 73
vacuum tubes 27
Van Allen, James 22
Vanguard TV-3 rocket 21
Velcro 146
vending machines 98
ventricular assist devices 133
Vickers VC10 82
Viking space probes 39
VisiCalc 53
volcanology 44-5, *86*
Voshkod 2 spacecraft *91*
Vostok 1 spacecraft 90

W

Walkman 29, 131
Walter, William Grey 37, *37*
Wang 52
Warhol, Andy 54
water jet cutters 87, *87*
Watt, William 46
weather forecasting 24-5
weather satellites 24, *24*, 27
weightlessness 91, 92, *92*, *93*, 97
Wensley, Roy 36, 37
Westinghouse Corporation 37
Wetzel, Donald C 98
Whinfield, Rex 108
White, Ed 76, *91*
Wichterle, Otto 99, *99*
Wiener, Norbert 37
Willie Vocalite (robot) 37
Windows 52
windsurfing 62, *62*
Wöhlk, Heinrich 99
Word 52, 53
word processing 52-3, *52, 53*
WordMaster 52
WordStar 52
World's Fairs 147
Wozniak, Steve 103-4
wristwatches 64, *64*, 109

X

X-ray computed tomography 116
Xerox 52, 63, 138
XV-8 Flexible Wing Aerial Utility Vehicle 33

Y

YAG lasers 30
Young, John 76
Young, Peter 153

Z

Zemdegs, Feliks 125
zip disks 69
Zweig, George 71

Picture credits

ABBREVIATIONS: t = top, c = centre, b = bottom, l = left, r = right

Front cover: main image, space walk, NASA. **Inset**: HP-65 calculator, Cosmos/SSPL/Science Museum. **Spine**: quartz watch, Seiko, France. **Back cover**: Concorde, Getty Images/BWP Media. **Page** 2, left to right, top row: Corbis/Bettmann; © Thinkstock 2010; BSIP/Photo Researchers/Bruce H Frisch; 2nd row: Roger-Viollet; BSIP/Photo Researchers/Scott Camazine; Seiko, France; 3rd row: Corbis/Bettmann; Cosmos/SPL/Georgia Lowell; Intel Museum/DR; bottom row: Getty Images/SSPL/Science Museum; © Thinkstock 2010; Cosmos/SSPL/Science Museum.

Pages 4/5: view of Earth taken from lunar orbit by Apollo 8 mission, Denis Cameron/Fex Features. 6l: Corbis/Bettmann; 6/7: Corbis/Science Faction/NASA; 7tl: Rue des Archives/RDA; 7b: Cosmos/SSPL/Science Museum; 7tr: Corbis/Gerolf Kalt; 8l: Computer History Museum, Mountain View, California; 8tr: Cosmos/SP/David Hardy; 8/9b: Corbis/Marc Garanger; 9tl: Cosmos/SPL/Dr L Caro; 9tr: Andia/Franck Godard; 9bc: BSIP/Photo Researchers/Bruce H Frisch; 10t: Cosmos/SPL/Fermilab; 10bl: Corbis/Smithsonian Institution; 10cr: Seiko, France; 11t: NASA/Johnson Space Center; 11bl: Corbis/Scott Speakers; 11br: NASA; 12t: Getty Images/Image Bank/Alain Choisnet; 12l: Intel Museum/DR; 12b: BSIP/Photo Researchers/Scott Camazine; 12/13: AKG Images/Ria Novosti; 13t: Merkur Museum, Czech Republic/Jan Suchy/DR; 13r: Corbis/Smithsonian Institution; 14tl: Cosmos/SPL; 14tr: Cosmos/SPL/Simon Fraser; 14b: Getty Images/SSPL/Science Museum; 15l: Cosmos/SPL/Pasieka; 15t: Getty Images/H Mark Weidman; 15c: Cosmos/SSPL/Science Museum; 16tl: REA/Elodie Grégoire; 16bl: Magnum Photos/Wayne Miller (detail); 16/17: Leemage/Delius; 17t: Cosmos/SSPL/Science Museum; 17cr: Cosmos/SPL/Pasieka; 17b: Library of Congress/Bill Graham Archives, LLC/The Art Archive/Lee Conklin; 18/19: NASA; 20bl: AKG Images/Ria Novosti; 20/21: AKG Images/Ria Novosti; 21tc&tr: Corbis/Bettmann; 22l: Corbis/Science Faction/NASA; 22tc: Corbis/Smithsonian Institution; 22br: AKG Images/Ria Novosti; 23bl: AFP Archives/RDA; 23tr: Cosmos/ESA; 24l: AFP/CNES; 24br: NASA; 25l: CNES/ESA/Arianespace/CSG Service Optique, 2007; 25tr: Cosmos/SPL/M-SAT Ltd; 26tl: AFP/CNES; 26bl: Corbis/Science Faction/NASA; 26br: Cosmos/SPL/NASA; 26/27: Corbis/Bettmann; 27cl: Cosmos/SPL/Rosenfeld Images; 27tr: Corbis/Bettmann; 27br: ESA/D Ducros; 28tl: Rue des Archives/RDA; 28bl: Gamma/Rapho/Keystone; 29tl: Corbis/Wendy Stone; 29tr: © Thinkstock 2010; 29b: © Thinkstock 2010; 30l: REA/Ian Hanning; 30br: Cosmos/SPL/Bo Veisland; 31t: Cosmos/SSPL/Science Museum; 31b: Cosmos/SPL/Tek Image; 32: Corbis/Bettmann; 33t: Corbis/Gerolf Kalt; 33bl: Corbis/Bettmann; 33br: NASA Image eXchange Collection; 34t: Getty Images/Jeffrey Phelps; 34b: Getty Images/Harrison Shull; 35t: REA/François Henry; 37b: © Noël Ansel, Pibrac;

36t: Corbis/Underwood&Underwood; 36b: Cosmos/SSPL/Science Museum; 37t: Corbis/John Springer/'The Day the Earth Stood Still', Robert Wise, 1951; 37b: Cosmos/SSPL/Science Museum; 38tl: Corbis/Bettmann; 38tr: Corbis/Charles O'Rear; 38b: Computer History Museum, Mountain View, California; 39b: NASA/STS-114 Crew; 39t: Corbis/Georgia Lowell; 40b: Corbis/EPA/Berdn Thissen; 40/41: Corbis/Gideon Mendel; 41b: 'Star Wars: The Phantom Menace', Georges Lucas, 1999, Lucas Films/Prod DB/DR; 42t: REA/Benoît Decout, 42b: © Pascal Goetgheluck; 43tl: Corbis/Ted Strehinsky; 43r: Corbis/EPA/Juerg Mueller; 44b: Corbis/Kipa/Jérome Minet; 44ht Corbis/Kipa/Jérome Minet; 45t: AKG Images/Alfio Garozzo; 45b: Cosmos/Xavier Marit; 46 & 46/47t: Cosmos/SSPL/Science Museum; 47cl: BSIP/Photo Researchers/Bruce H Frisch; 47b: AFP/Gary M Prior; 48l: Cosmos/SSPL/Science Museum/Jennie Hills; 48b: AFP/Ferrari Press Office; 49b: Cosmos/SPL/NASA; 49t: Corbis/David Brabyn; 50: Rue des Archives/AGIP; 51bl: Rue des Archives/AGIP; 51tr: Cosmos/SPL/Dr L Caro; 51br: Cosmos/SPL/Elena Kiseleva; 52: Corbis/ClassicStock/K Vreeland; 53tl: Logiciel WORD/DR; 53c: Getty Images/Time Life/Steve Liss; 54t: Collection Centre Pompidou, Dist RMN/ADAGP, Paris 2010/'Soudain l'été dernier' Martial Raysse, 1963; 54b: Corbis/Moodboard; 55t: Roger-Viollet; 55b: Corbis/Chris Caroll; 56tl: AFP/Marcel Mochet; 56r: Hemis.FR/Bertrand Rieger; 57t: Andia/Franck Godard; 57b: Rue des Archives/AGIP; 58t: Corbis/Marc Garanger; 58r: AKG Images; 59b: Corbis/Richard Hamilton Smith; 59tr: Cosmos/Pascal Maitre; 60l: Cosmos/Jonathan Blair; 60tr: Cosmos/SPL/David Hardy; 61t: Cosmos/SPL/Julian Baum; 61cl: Cosmos/SPL/NASA; 62tr: Corbis/Rick Doyle; 62b: Getty Images/Norbert Eisele Hein; 63t: SRI International, Menlo Park, California; 63b: REA/HH/Martin Van de Griendt; 64: Seiko, France; 65t: Cosmos/SPL/James Bell; 65bl: Cosmos/SPL/Manfred Kage; 66b: REA/Alexandre Gelabart; 66t: REA/Redux/Tom Wagner; 67l: Corbis/Bettmann; 67t: Corbis/Gene Blevins; 68bl: Corbis/Sygma/Frederic Neema; 68tr: Reuters/Pool/Rick Rycroft; 69bl: Corbis/Scott Speakers; 69tr: Corbis/Science Faction/Stefan Sollfors; 70bl: Cosmos/SPL/Fermilab; 70tr: © Daniel Bonnerue; 71tl: Getty Images/Hulton Archives/Photoshot; 71r: Courtesy Peter Ginter/SLAC National Accelerator Laboratory; 72t: Roger-Viollet/TopFoto; 72b: AKG Images/SPL; 73t: AKG Images/Ullstein bild; 73b: Corbis/Science Faction/NASA; 74l: NASA, Marshall Space Flight Center Collection; 74cr: NASA; 75b: Corbis/Bettmann; 75cr, 76t 76b, 77t, 77b, 78l, 78/79, 78br & 79tr: NASA; 80tl: Getty Images/Business Wire; 80cr: Getty Images/Realistic Reflexions; 81tr: BSIP/Photo Researchers/Scott Camazine; 81b: REA/Laurent Cerino; 82tl: Corbis/Paul Almasy; 82b: AKG Images/Ria Novosti; 83t: Getty Images/Image Bank/Alain Choisnet; 83bl: Corbis/Jack Fields; 83cr: Corbis/Bettmann; 84t: Getty Images/BWP Media; 84b: Corbis/Sygma/Maurice Rougemont; 85t: REA/LAI/Martin Sasse; 85cr: Corbis/SYGMA/Bernard Annebique; 86t: Corbis/Russ Schleipman;

86bl: Corbis/Georges Steimets; 86br: Corbis/Paul Souders; 87c: REA/Didier Maillac; 87tr: Cosmos/SPL/Georgia Lowell; 88tr: Corbis/Roger Du Buisson; 88r: Intel Museum/DR; 89t: Corbis/Construction Photography; 89b: Intel Museum/DR; 90c: NASA, Marshall Space Flight Center Collection; 90b: AKG Images/Ria Novosti; 90/91: Corbis/Smithsonian Institution; 91r: AKG Images/Ria Novosti; 92t: NASA; 92l: AFP/The Coca Cola Company; 93t: Corbis/Roger Ressmeyer; 93bl: Corbis/Roger Ressmeyer; 93r: NASA/Johnson Space Center; 94: AKG Images/Ria Novosti; 95tl & 95r: NASA; 95br: AKG Images/Ria Novosti; 96t: Corbis/EPA/Yuri Kotchetkov; 96c: NASA; 96bl: Reuters/Mark Baker; 96/97t: NASA; 97tr: NASA Image Spaceflight Collection; 98bl: AKG Images/Theodor Braun, Vienna, 1965, Gauselmann Collection; 98tr: 'L'Œil du mal', ('Eagle Eye'), D J Caruso, 2008, Prod DB/DreamWorks SKG – Goldcrest Pictures/DR; 99tl: REA/Michel Gaillard; 99r: Merkur Museum, Czech Republic/Jan Suchy/DR; 100t: Getty Images/Ryan Creary; 100b: Corbis/Karl Weatherly; 101: X Duret/B Boone; 102l: Corbis/Bettmann; 102r: Cosmos/SSPL/Science Museum; 103t: Computer History Museum, Mountain View, California/Gift of Armand Van Dormael; 103b: Cosmos/SPL/Volker Steger; 104l: Cosmos/SSPL/Science Museum; 104t: Getty Images/SSPL/Science Museum; 104b: Corbis/Roger Ressmeyer; 105bl: Photos12.com/Alamy/Sergey Galushko; 105br: REA/Benoît Decout; 106t: Musée de l'Informatique, Paris, www.MuseeInformatique.fr/Guillaume Beguin; 106bl: Cosmos/ClassicStock/D&P Valenti; 107cl: Getty Images; 107t: REA/Richard Damoret; 108c: Corbis/Stefano Bianchetti; 108/109: REA/Hamilton; 109tr: © Thinkstock 2010; 109cr: Leemage/Stefan; 110t: **REA/Stern-Laif/Rethmann AG**; 110b: Hemis.FR/Bruno Morandi; 111t: Getty Images; 111c: REA/Berti Hanna; 111b: Cosmos/SPL/James King-Holmes; 112b: Reuters/David Gray; 112t: Cosmos/SSPL/Science Museum; 113tl: REA/Antoine Devouard; 113r: Cosmos/SPL; 114l: Corbis/Papilio; 114/115: Corbis/Eric Nguyen; 115r: Cosmos/SSPL/Pasieka; 116l: Cosmos/SPL; 116r: Cosmos/SSPL/Science Museum; 117t: Corbis/Bettmann; 117cr: Cosmos/SPL/Simon Fraser; 117b: REA/Ian Hanning; 118: Corbis/San Francisco Chronicle/Michael Maloney; 118bl: Getty Images; 118br: Getty Images/Jay Colton; 119: Getty Images/H Mark Weidman; 119b: Corbis/Car Culture; 120b: REA/Laurent Grandguillot; 120/121: Leemage/Delius; 121tr: REA/James Leynse; 121cr: Getty Images/Time Life/Bill Bridges; 122b: REA/Financial Times/C Bibby; 122t: Corbis/Benelux; 123t: REA/Laif/Joerg Glaescher; 123b: REA/Gilles Rolle; 124t: Corbis/Aurora Photos/Chris Milliman; 124b: REA/Richard Damoret; 125c: Getty Images/AFP 2007/Attila Kisbenedek; 125br: REA/Didier Maillac; 126t: Cosmos/SPL/Cordelia Molloy; 126b: Getty Images/National Geographic/Joseph D Lavenburg; 127: Corbis/Sygma/Alain Nogues; 128t: REA/Gilles Rolle; 128bl: Corbis/Tom Stewart; 128br: REA/Elodie Grégoire; 129tl: REA/François Perri; 129br: Corbis/Masterfile/

Radius Images/Edward Pond; 130/131t: Corbis/Duncan Smith; 130c: Corbis/Bettmann; 130/131b: REA/Laif/Thomas Coste; 131br: Corbis/Blend Images/Granger Wootz; 132t: AFP/Lionel Bonavanture; 132b: Cosmos/SSPL/Science Museum; 133t: Getty Images/SSPL/Science Museum; 133cr: Cosmos/SPL/Eye Of Science; 133b: Cosmos/SPL/Coneyl Jay; 134l: AFP/MFP/Michel Brunet, 134/135: Corbis/Frans Lanting; 135tl: Corbis/Christophe Boisvieux; 135tr: Musée Préhistorama de Bidon, Ardèche/AFP; 136: AFP/Eric Feferberg; 137t: Corbis/Olivier Cadeau; 137b: Cosmos/SPL/Pasieka; 138tl: Magnum Photos/Wayne Miller (detail); 138bl: Corbis/Kim Kulish; 139t: Corbis/Charles O'Rear; 140c: AKG Images; 140b: Hemis.FR/Pawel Wysocki; 140/141t: Hemis.FR/Pacific Stock; 141bl: Magnum Photos/Elie Reed (detail); 141tr: Library of Congress/Bill Graham Archives, LLC/The Art Archive/Lee Conklin; 144bl: AP/Press Association Images; 142tr: Corbis/Henry Diltz; 143t: AKG Images/Hans W Mende; 143t: Corbis/Ted Soqui Photography 2009; 144/145: Cosmos/SPL/James Bell; 146bl: AKG Images/Ria Novosti; 146tr: Corbis/Wendy Stone; 146br: © Thinkstock 2010; 147bl: NASA, Marshall Space Flight Center Collection; 147bc: Cosmos/SSPL/Science Museum; 147r: Corbis/Bettmann; 148bl: AKG Images/Ria Novosti; 148t: Roger-Viollet; 148br: Cosmos/SPL/Manfred Kage; 149bl: Corbis/Richard Hamilton Smith; 149r: NASA; 150bl: Cosmos/SPL/Georgia Lowell; 150c: 'L'Œil du mal', ('Eagle Eye'), D J Caruso, 2008, Prod DB/DreamWorks SKG – Goldcrest Pictures/DR; 150br: Corbis/Karl Weatherly; 151t: Corbis/Aurora Photos/Chris Milliman; 151bl: Cosmos/SSPL/Science Museum; 151bc: Corbis/Christophe Boisvieux; 151br: Musée Préhistorama de Bidon, Ardèche/AFP; 152l: Cosmos/SPL/Julian Baum; 152tr: Cosmos/SPL/Cordelia Molloy; 152br: Corbis/Kipa/Jérome Minet; 153l: Cosmos/SSPL/Pasieka; 153r: Rue des Archives/AGIP.

Illustrations on page 71 (the components of matter) by Grégoire Cirade.

THE ADVENTURE OF DISCOVERIES AND INVENTIONS
Into the Space Age – 1960 to 1975
Published in 2011 in the United Kingdom by Vivat Direct Limited
(t/a Reader's Digest), 157 Edgware Road, London W2 2HR

Into the Space Age – 1960 to 1975 is owned and under licence from
The Reader's Digest Association, Inc. All rights reserved.

Copyright © 2011 The Reader's Digest Association, Inc.
Copyright © 2011 The Reader's Digest Association Far East Limited
Philippines Copyright © 2011 The Reader's Digest Association Far East Limited
Copyright © 2011 The Reader's Digest (Australia) Pty Limited
Copyright © 2011 The Reader's Digest India Pvt Limited
Copyright © 2011 The Reader's Digest Asia Pvt Limited

Adapted from *L'Ère de la Conquête Spatiale*, part of a series entitled L'ÉPOPÉE DES DÉCOUVERTES ET DES INVENTIONS, created in France by BOOKMAKER and first published by Sélection du Reader's Digest, Paris, in 2011.

Reader's Digest is a trademark owned and under licence from The Reader's Digest Association, Inc. and is registered with the United States Patent and Trademark Office and in other countries throughout the world. All rights reserved.

All rights reserved. No part of this book may be reproduced, stored in a retrieval system, or transmitted in any form or by any means, electronic, electrostatic, magnetic tape, mechanical, photocopying, recording or otherwise, without permission in writing from the publishers.

Translated from French by Tony Allan

PROJECT TEAM
Series editor Christine Noble
Art editor Julie Bennett
Designers Martin Bennett, Simon Webb
Consultant Ruth Binney
Proofreader Ron Pankhurst
Indexer Marie Lorimer

Colour origination FMG
Printed and bound in China

VIVAT DIRECT
Editorial director Julian Browne
Art director Anne-Marie Bulat
Managing editor Nina Hathway
Picture resource manager Sarah Stewart-Richardson
Technical account manager Dean Russell
Product production manager Claudette Bramble
Production controller Sandra Fuller

We are committed both to the quality of our products and the service we provide to our customers. We value your comments, so please feel free to contact us on 0871 3511000 or via our website at **www.readersdigest.co.uk**

If you have any comments or suggestions about the content of our books, you can email us at **gbeditorial@readersdigest.co.uk**

CONCEPT CODE: FR0104/IC/S
BOOK CODE: 642-012 UP0000-1
ISBN: 978-0-276-44524-8